MANAGING SERVICE LEVEL QUALITY
ACROSS WIRELESS AND FIXED NETWORKS

MANAGING SERVICE LEVEL QUALITY
ACROSS WIRELESS AND FIXED NETWORKS

Peter Massam
Hutchison 3G UK Ltd

JOHN WILEY & SONS, LTD

Telephone (+44) 1243 779777

Email (for orders and customer service enquiries): cs-books@wiley.co.uk
Visit our Home Page on www.wileyeurope.com or www.wiley.com

Other Wiley Editorial Offices

John Wiley & Sons Inc., 111 River Street,
Hoboken, NJ 07030, USA

Jossey-Bass, 989 Market Street,
San Francisco, CA 94103–1741, USA

Wiley-VCH Verlag GmbH, Boschstr. 12,
D-69469 Weinheim, Germany

John Wiley & Sons Australia Ltd, 33 Park Road,
Milton, Queensland 4064, Australia

John Wiley & Sons (Asia) Pte Ltd, 2 Clementi Loop #02-01,
Jin Xing Distripark, Singapore 129809

John Wiley & Sons Canada Ltd, 22 Worcester Road,
Etobicoke, Ontario, Canada M9W 1L1

British Library Cataloguing in Publication Data

A catalogue record for this book is available from the British Library

ISBN 0470 84848 0

Typeset in 10/12pt Times by Deerpark Publishing Services Ltd, Shannon, Ireland

This book is printed on acid-free paper responsibly manufactured from sustainable forestry in which at least two trees are planted for each one used for paper production.

Contents

Preface

In 1998, as part of a global deployment of a mission-critical monster of an application, I was tasked with a brief to guarantee that the mythical source of all performance-related problems, which has come to be known as 'the network', was not to be blamed for degradations experienced by the users of that application. At least, if it were to blame, we should be able to produce the hard evidence to substantiate the claims and provide corrective actions quickly and effectively.

On the face of it, this would appear to be a reasonable and simple request, but isolating the network portion from the rest of the transaction does not give the full picture. Comforting though it may be to know that your part of the network is functioning correctly, it does nothing to identify the source of performance problems, but rather remains a source of constant irritation to business and consumer customers alike. Like the Emily Pankhursts of their day, customers of these applications have risen up to demand what was rightfully theirs – a decent level of service.

Embodied in that right are many different interpretations. For some customers, it may be working at the speed you and your thought processes want to, for others, it would be working anywhere at any time at the speed you want to. In both cases, this really boils down to how responsive a function or an application is. Put more simply, when you press the button or key, how long do you have to wait to get a response.

It was clear that a solution had to be found that encompassed all portions of a transaction from key press through the network to the server and back again, and that slowdowns and their sources were flagged up at the earliest opportunity before customers were impacted. In the light of this, a service-level management strategy was devised, planned, and implemented globally within 12 months.

It is that experience that I wished to share, as it became obvious that not only was there a good deal of interest from both internal and external customers in that strategy, but also, one of the fringe benefits, namely truly proactive end-to-end management, was being realised. That account with an example of the practical implementation of such a strategy occupies the first half of this book.

The second half reflects both the movement in the industry towards mobile computing and the particular challenges when trying to apply a similar strategy across fixed and wireless networks. Here, the approach is one of familiarising yourself with wireless environments, understanding what is going to help you maintain service levels, and providing some indicators as to the expectations you need to set when passing what have become familiar applications over the air to a mobile device.

To understand these challenges, it is important to understand the technologies, what lies behind them, and what mechanisms are available to assist you in capturing the all-important real-time information on application performance, which will convey to you the pain as well as the pleasure that your customers are feeling.

Both halves cover aspects of technologies that relate only to application performance. It is not meant to be a thorough examination of SNMP, MIB structures, or RMON groups, which have been adequately covered by many acknowledged in the references section. What it does give is an insight into why service levels are important, how to implement a service-level management strategy based on application performance and what performance is likely to be like in a wireless environment.

The test results in the concluding section of this book are meant to promote further investigation by those interested in the area of application performance.

This was written both for IT professionals, from either a wireless or an IP data networking background who wish to familiarise themselves with the other half of the equation, and for anyone with an interest in how service quality is maintained and delivered across different types of networks.

Armed with the information contained herein, I hope that you will feel better equipped to deal with the challenge that bringing fixed and wireless worlds together presents and that it sets you well on the road to delivering a service quality that your customers will appreciate and that you can be proud of.

Colour versions of the figures contained within the book are available at the following URL: ftp://ftp.wiley.co.uk/pub/books/massam

List of Figures

List of Test Figures

List of Tables

Acknowledgments

The author wishes to acknowledge the continued unerring support and encouragement of my wife, Fiona, and the distractions and light relief that are our children, William and Francesca.

Introduction

Understanding applications' performance is to know the frustrations felt every day by business and consumer customers alike. Without it, we divorce ourselves from the reality of slowdowns or 'performance brownouts', which undoubtedly cost businesses a tidy sum in lost time and inefficiencies. For those whose job it is to look after such customers, it can be a lifeline to find a means of gleaning real-time application response time information to aid in root-cause analysis before a business-impacting event takes place.

Added to this, there is an increasing expectation that not only must the network deliver on its 99.999% availability promises, but that applications' response time should also be an intrinsic part of any Service Level Agreement (SLA) between service provider and customer.

In recent times, the technology industry has been promoting the 'anytime, anywhere' concept of the all-pervasive Internet by talking up the ability to deliver all services to any handheld Internet Protocol (IP) device across the airwaves and is now being gripped by the grim reality of having to invest heavily for the privilege. Much of the fervour in 2000 centred around the auction of licences for the third-generation (3G) mobile networks and the preparation of applications to run over them, but we also saw the relaxing of restrictions in summer 2001 allowing access to wireless Local Area Networks (LANs) in public places, which has done much to bolster a flagging LAN switch market desperately looking for new products to justify new price tariffs.

While vendors vie for position over the ultimate access device, developers and Internet Service Providers (ISPs) promote their portal as the best way into the Internet, and standards bodies endeavour to keep up, it is easy to lose sight of a basic prerequisite to this activity: will it work?

We begin with an introduction to real-time applications performance as a means of delivering service level management (SLM) and of aiding root-cause analysis in fixed networks. We then go on to examine what it takes to *guarantee* service-level quality across fixed and wireless networks today, what is on offer tomorrow, what architectures have to be accommodated in this new, dare we say, converged model, and what levels of performance we can expect from it.

As with many new concepts, terminology can be misleading. So, for the purposes of this book, the term 'multiservice networks' simply alludes to networks that carry multiple services namely voice, data, and video.

1

Managing Service Level Quality in Fixed Networks

We begin the search by examining what mechanisms are available to provide service-level management in fixed or wireline networks today. To manage the 'customer experience' effectively, it is necessary to consider the real-time monitoring of applications' response times to feel the pain and sometimes the pleasure that a customer feels, to examine what future standards may have to offer, and to determine a viable method of delivering response time information that will aid root cause analysis.

Many mainstream network management applications claim to be 'enterprise-wide' and 'end-to-end applications monitors', but when examined more closely, few are scalable for large global enterprises, and none give a full view from client through to the server.

1.1 Relevance

In order to minimize the number of applications that support personnel would need to look at to perform a root cause analysis, this document examines what standards or structures support applications response time monitoring already and endeavours to find a working solution.

1.2 Scope

The following areas are covered:

- an historical look at management methods and the limited services they provide;
- an assessment of current standards and those drafted for consideration;
- the latest developments in the use of intelligent agents;
- the merits of active and passive monitoring;
- a practical integration model to maximize benefits of proactive monitoring; and
- recommendations on best practice and implementation of such a model.

1.3 Summary

There is no one tool that will give the 'drill down' functionality required to perform a root cause analysis. The key is integration of a minimal tool set with minimum administrative overhead, which delivers real-time information to the people (Operational Systems Support, for the most part) who need it, and one that reports by exception. This is achievable today and has been implemented globally on the corporate network of a large supplier to the telecommunications industry.

2

History

This chapter is intended to give an overview of the traditional functional areas associated with Network Management[1] and to examine the established sets of standards that have been applied in the search for one that delivers real-time monitoring of applications' response times.

In turn, the following are examined:

- Simple Network Management Protocol (SNMP [1] versions 1, 2 and 3);
- Management Information Base (MIB I, MIBII [5] and extensions);
- Remote Network Monitoring (RMON and RMON2);
- Common Management Information Services (CMIS).

2.1 Summary

Various attempts have been made to standardize on a single protocol, but all have fallen into disuse largely because of their over-complexity. RMON 2 represents the only attempt to categorize the characteristics of applications traffic and breaks it down by protocol.

While these standards give an idea of the amount of traffic taken up by an application and can identify which computer system is responsible for it and to which other system it is talking, the RMON2 specification does not calculate application response time, and some proprietary extensions are limited to Round Trip Time (RTT) measurements based on delta values of observed traffic.

With the advent of merged data, voice, and video networks or Unified Networks[2], it has become necessary to classify types of traffic to guarantee timely delivery of packets either in an ordered or regular pattern (as with passing voice or real-time video over data packet networks) or within a given time frame because of the time sensitivity of the application or because of the degradation of quality that would result (as with non-real-time video).

Packet InterNet Groper (PING) has been a tool traditionally used to test response times, but this relates only to its own protocol (ICMP) and has been proven to be woefully inadequate when dealing with applications traffic (see Appendix A).

---

[1] Network Management is taken here to include voice, video, and data networks, computing systems, and related data-transmission and data-processing devices.

[2] Unified Networks is a trademark of Nortel Networks.

2.2 Traditional Network Management

One cannot start looking at existing standards of Network Management without first understanding what areas they are attempting to address. This section looks at the five key areas, the automated tools required at a basic level, and a comparison of surveys conducted in 1992 and 2000 on the importance placed by users on features [6, p. 3].

2.2.1 OSI Functional Areas

The ISO developed a functional breakdown to turn user requirements into a structure capable of managing OSI systems but can equally be used to define the requirements of any network management system (NMS). The five key areas are:

- Fault Management;
- Performance Management;
- Configuration Management;
- Security Management;
- Accounting Management.

Expanded definitions of the categories above are listed in Table 2.1 [6, p. 4]:

Table 2.1 OSI functional areas

OSI management functional areas	Definition
Fault Management	Enables the detection, isolation, and correction of abnormal operation of the OSI environment
Performance Management	Evaluates the behaviour of managed objects and effectiveness of communications activities
Configuration Management	Exercises controls over, identify, collect data from and provide data to managed objects to assist in continuous operation of interconnection services
Security Management	Implements OSI security model correctly and protects managed objects
Accounting Management	Enables charges to be identified for the use of managed objects

By and large, vendors of NMSs tend to concentrate on one or all of the first three above and give little or no consideration to the rest. Nowhere here do we find any definition for Applications Management, which was not a priority at the time this model was defined.

2.2.2 Network Management Features Deemed Most Important (1992)

It is interesting to note how requirements have changed since this survey was conducted in 1992. Figure 2.1 puts monitoring response time firmly last.

While ease of use would still be top of the list given the overall scope of the task any NMS is being asked to do, the emphasis eight years on is much more on Service Level

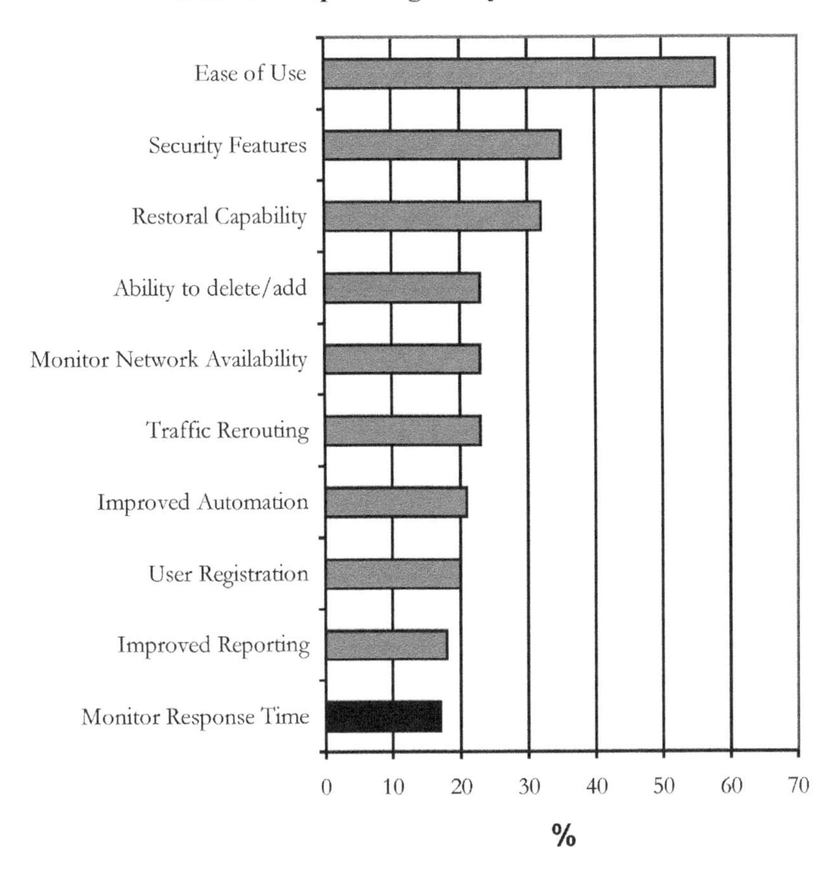

Figure 2.1 Most important network management features (1992). Source: International Data Corp. (1992).

Agreements (SLAs) and the data it is necessary to collect in order to support them. This is discussed in more depth in Chapter 3 (Section 3.4 The SLA Factor). Now, response time is divided into subcategories and is ranked much higher on the list of desirable features, as shown below.

2.2.3 Network Management Features Deemed Most Important (2000)

The change in emphasis can be seen clearly in Figure 2.2 with this study taken in May 2000. Monitoring response time has now been broken down into its relevant component parts: those of 'application performance management' and 'network-wide performance monitoring and health indicators', accounting for 57% and 34%, respectively.

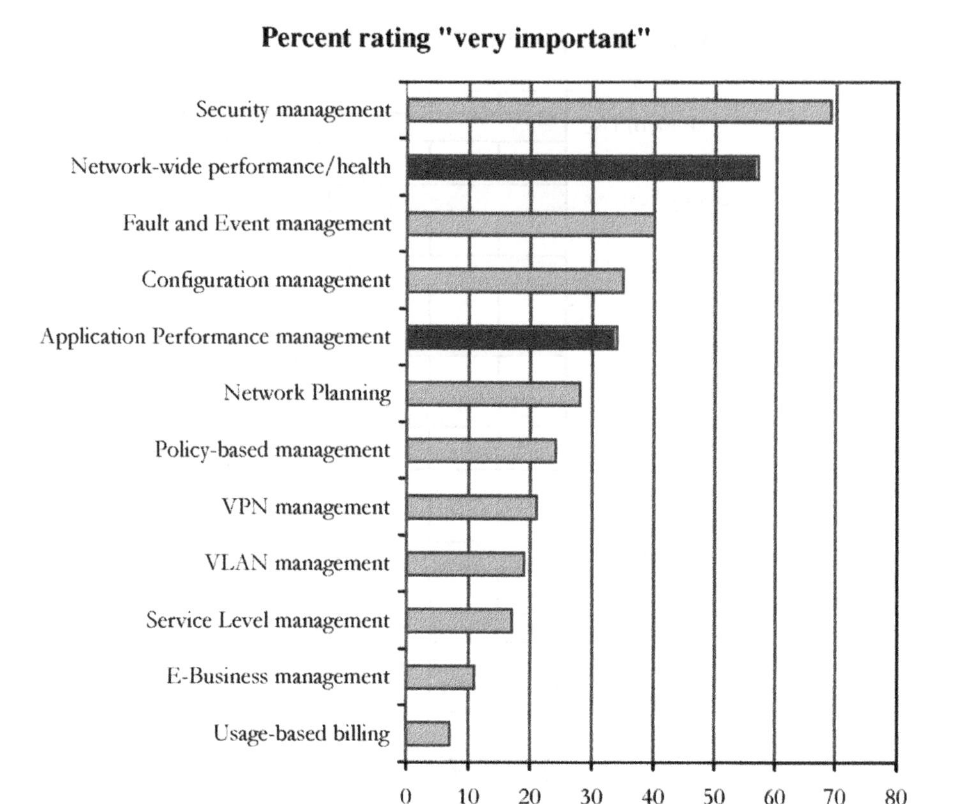

Figure 2.2 Most important network management features (2000). Source: Infonetics Research *User Plans for Network Management* (2000).

2.2.4 Automated Tools Requirement

Stallings [6] states that any NMS should enable users to do the following:

- Configure – set up components, initialize subsystems, and set parameters;
- Operate – monitor utilization and perform accounting functions;
- Maintain – react to failures and overloads;
- Control Access – define and enforce prioritized security access controls;
- Plan – capacity plan for additional services and growth.

These remain essential to maintaining a network, but again, security tends to be limited to the security on the Network Management Station itself with access being granted to it by configurations on the network devices.

Capacity planning in data networks remains the preserve of a few niche vendors who provide vendor-specific, usually not inexpensive tools.

Telecommunications network management tools, however, have always had an element of capacity planning built in to determine scientifically when growth is about to exceed capacity.

Data network modelling tools conversely are complex in nature and struggle to store enough data from a live network to provide scientific, projected forecast analysis.

An abstract [7] that still holds true today is one that lists the challenges of managing Local Area Networks and Wide Area Networks (LAN/WAN) as follows:

> ....collection...of management information from diverse network elements, applying local and network-wide intelligence for correlating network events and isolating problems, and ... creating a flexible management structure which reflects the growing and changing organizational needs. ...the architecture should be expandable for ...increasingly large sized networks and must not allow the management information to impose an excessive overhead on the network resources.

This introduces the important element of time when 'correlating' activities. This is a subject I will return to in Section 2.4. It is a function of the NMS to correlate such events and not the remote devices.

For now, it should be noted that SNMP has only a 'loose' mechanism for time synchronization[3], and it is very much left to the NMS to make a valued judgement on both when a message was sent and whether, by inference, it was from an authentic source.

The rest of this chapter is devoted to the types of 'management information' that have been available and their flexibility, expandability, and suitability, or not, for handling application performance.

2.3 Simple Network Management Protocol (SNMP)

This section identifies the origins of SNMP, what is contained in the set of standards an d whether subsequent versions have moved on sufficiently to deliver real-time monitoring of applications' response times.

The nature of how we monitor networks has changed dramatically over the last decade.

From a standing start, SNMP has dominated the market because of its simplicity and flexibility. Vendors are free to develop their own extensions to the basic requirements and have delivered custom applications to make best use of these extensions.

SNMP version 2c is an agreed standard introducing some long-overdue security features, but these were not as comprehensive as some would have liked, and SNMPv2 is only beginning to gain acceptance among vendors. Version 3 has yet to appear on most vendors' radars.

Attempts to address security measures or to migrate wholesale to the OSI 'long term' solution (CMIS) were resisted by most vendors, who viewed their structures as over complex and requiring too much resource for very little return on investment (ROI).

2.3.1 SNMP Origins

SNMP [1] was born out of the need to monitor IP gateways on the Internet. The forerunner that was called Simple Gateway Monitoring Protocol [2] was designed with minimum features. It did not include authentication *per se*, as control of those gateways was not a priority at that point. However, hooks were there for the later addition of this functionality.

---

[3] SNMP was built not to rely on other network services such as *Network Time Protocol*(NTP) or secret/key management protocols.

SNMP was only ever meant to be a short-term standard [3], making way for the full-blown CMIS, which forms part of the Open Systems Interconnection (OSI) framework for managing and monitoring OSI networks using International Standards Organization (ISO) protocols.

As any Network Manager will testify, many short-term fixes become long-term solutions without really trying. This is usually down to a combination of their simplicity of construction, low maintenance and seemingly endless sustainability.

2.3.2 SNMP (Renamed SNMP v1)

This section briefly describes SNMP and gives an understanding of 'traps', essential when attempting to combine real-time monitoring with root cause analysis.

SNMP is part of the Transmission Control Protocol/Internet Protocol (TCP/IP) suite and as such was developed to monitor IP networks.

It is not one but a *set of standards* incorporating a set of objects called the Management Information Base (MIB), a database structure known as Structure of Management Information (SMI) [4], and a protocol to run them over which is SNMP. SNMP is a protocol allowing an SNMP Network Management Station (NMS) to communicate with an SNMP agent using five basic commands to retrieve and set parameters.

An NMS issues *GetRequest, GetNextRequest,*and *SetRequest*commands to retrieve or set parameters, which are returned via a *GetResponse* from the SNMP agent. The agent can also warn the NMS of potentially calamitous events by issuing a *trap*, which is an unsolicited message. If the message contents are vendor or enterprise-specific, the message can be interpreted only if the NMS has the appropriate MIB information loaded for that vendor's device.

Figure 2.3 refers to these basic communication flows.

2.3.2.1 SNMP Traps

The scenario in Figure 2.3 only shows three devices enabled with SNMP. In a large network of hundreds or thousands of routers, switches, and hubs, an architecture is needed to support that kind of infrastructure. Some hints are given at the end of this document suggesting ways to effectively manage your management traffic! (See Chapter 6: An Integration Model).

A *trap* is an unsolicited message sent from an SNMP agent to any NMS it has been configured to send it to. The content of these traps ranges from a 'simple' interface status (up or down) to power and environmental information.

An advantage of enabling traps from an SNMP-enabled device is that one can be warned of a fault or impending fault immediately without waiting for the NMS to cycle round to asking that specific device for information. This can also reduce the amount of network management traffic travelling over expensive WAN links. However, read on for an important caveat below.

One inherent danger of permitting unsolicited traps to be generated is the flooding of information to an NMS in the event of a serious problem. Not only can this slow down the performance of the NMS to a standstill, but, more importantly, it can conceal the very piece of information that would lead to the root cause in a morass of minutiae, which scrolls so quickly past that it goes unnoticed.

Controlled traps are, however, an important part of any corporate NMS and, when properly used, can be highly effective in tracking down root causes, even at applications level.

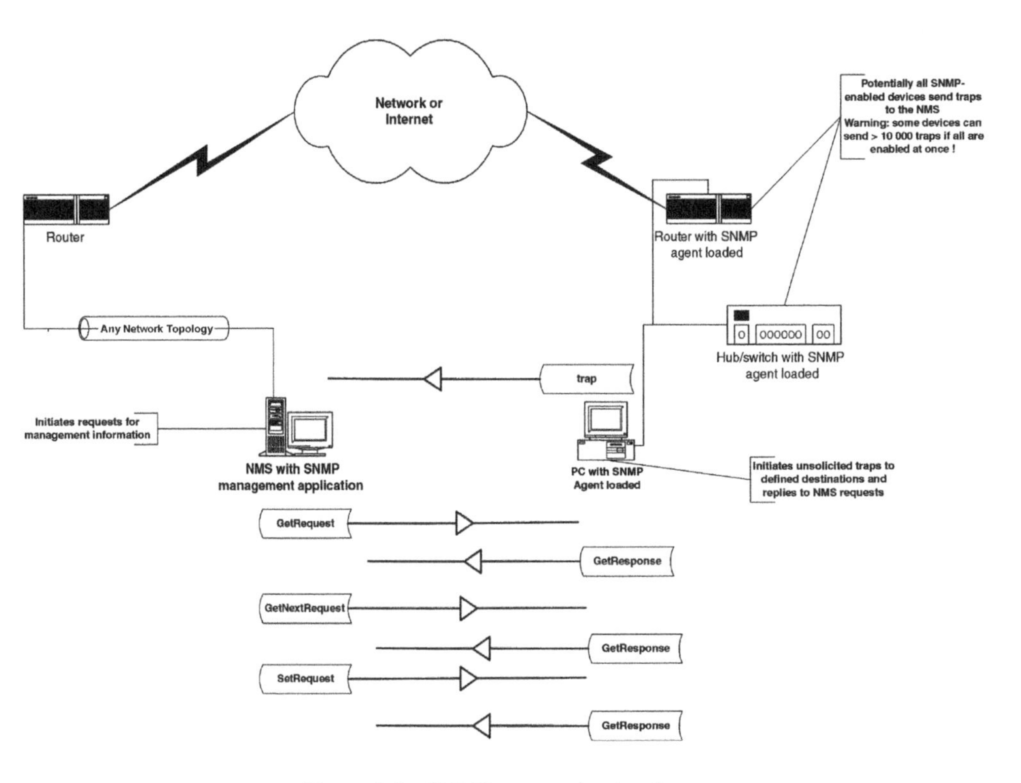

Figure 2.3 SNMP communication flows.

One thing should be noted at this point: a *trap* only becomes an *event* after meaningful words have been written to describe the trap, and a severity level (critical, major, minor, warning or normal) has been assigned to it. This is a *manual* process.

2.3.3 SNMP Version 2 (SNMP v2c)

This section describes the enhancements made to SNMP and any impact they had on monitoring real-time applications response time. The three areas that were affected by SNMP v2 were:

- Structure of Management Information (SMI) modifications;
- Protocol operations and interoperability with SNMPv1;
- Manager-to-manager communications.

Each of these enhancements, although seemingly small, was necessary to expand the role of SNMP in a secure way and made it accessible to distributed network management systems instead of just serving highly centralized systems.

2.3.3.1 SMI Modifications

SMI (RFC1155) refers to a structure with which MIBs and their objects can be defined and specifically lays down the syntax to be used, the data types, and the relationships to other MIB objects.

The syntax used is Abstract Syntax Notation (ASN.1), which is of the form:

1.3.6.1.2.1..., with each number representing a path of a directory tree. The first four numbers remain the same, as these refer to the 'Internet' branch beneath which all other branches lie. See Section 2.4 for more details on the MIB layout.

Several new data types were added under SNMPv2, and documentation associated with objects was enhanced to make more sense of the information that was being passed.

Two more access categories were added: *read-create* and *accessible-for-notify*, the latter being used in the Manager-to-manager communications, and the *write-only* was dropped.

Also, one could see basic traffic associated with the SNMPv2 protocol as well as agent and manager configuration through an additional SNMPv2 MIB.

2.3.3.2 Protocol Operations and Interoperability with SNMPv1

The main change to SNMPv2 was the addition of two more commands:

- *GetBulkRequest*– avoided multiple serial requests of the *GetNextRequest* statement of SNMPv1;
- *InformRequest*– enabled communication of trap information from one manager to another.

This promoted efficient transport of bulk data reducing the need for multiple small packets of information to be transmitted and extended the scope of the NMS role to be able to forward trap information on to each other.

Interoperability with the previous version is achieved through one of two methods:

- Interoperability Method1: An SNMPv2 agent can act as a proxy agent for any SNMPv1 devices, converting between the two formats. Principally, this means mapping the *GetBulkRequest* to a *GetNextRequest* and converting traps from SNMPv2 format to version 1.
- Interoperability Method2: An SNMP manager (NMS) can speak in and understand both formats. This is often referred to as a 'bilingual manager'.

2.3.3.3 Manager-to-manager Communications

The important change here is that one NMS acting as an SNMP manager can request SNMP information about devices that another NMS is looking after as an SNMP manager in its own right. This effectively allows propagation of trap information and MIB information through the network between managers, allowing more complex NMS architectures to be built.

Figure 2.4 highlights the changes in the communication flows for SNMPv2.

Although not supplying information directly related to applications or their response times, SNMPv2 does supply some of the mechanisms for passing information between SNMP managers (NMS), which will be needed to monitor such response times in real-time.

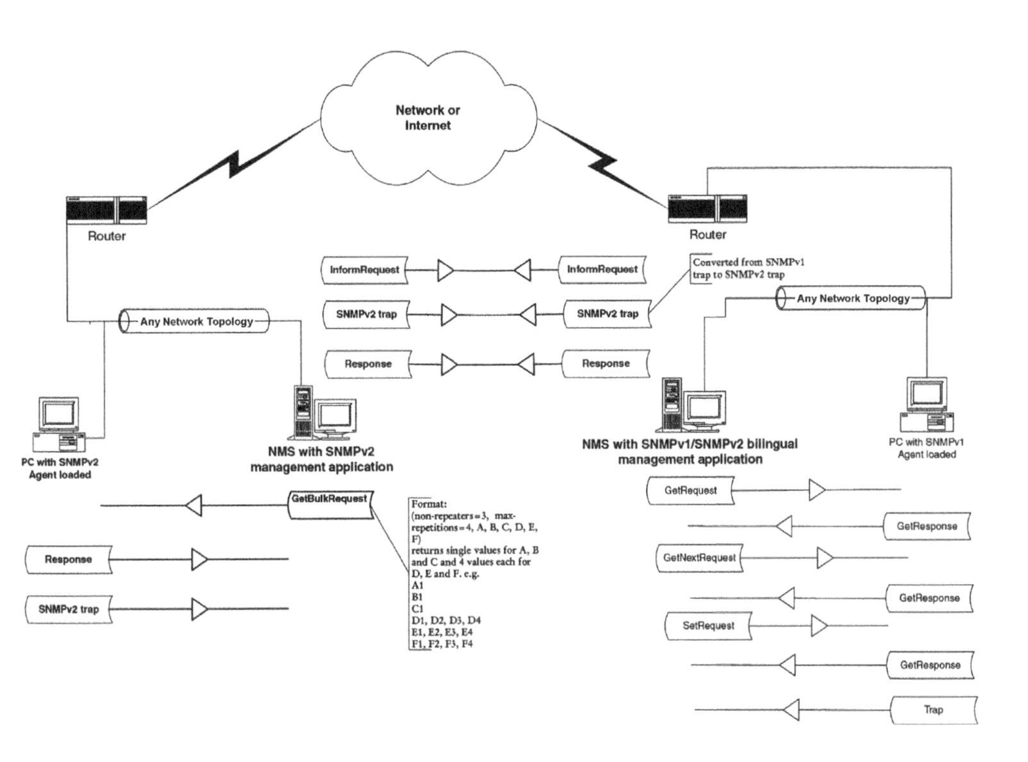

Figure 2.4 SNMPv2 communication flows.

2.3.4 SNMP Version 3 (SNMPv3)

This section outlines the most recent version of SNMP, what it achieves, and any impact it may have in assisting in the delivery of response time information.

It must be stated that this is not a new standard but simply supplies a secure and future-proof structure for the two existing standards to operate in. One such future proofing is the inclusion of the 16 octet IPv6 addressing in the schema.

In this respect, the ways in which SNMP v1 and v2 contribute to supply monitoring data from managed agents hold true, therefore, for SNMPv3.

SNMPv3 is not a stand-alone standard. Specifically, it defines the following:

- an SNMP engine – a set of four subsystems for processing messages;
- SNMP applications – a set of five for generating and receiving messages;
- two SNMP Security models:
  - User-based Security Model (USM)
  - View-based Access Control Model (VACM).

The following subsections describe briefly the functions of each of these.

2.3.4.1 SNMP Engines

Note that the traditional agent and manager categorizations have now been turned into SNMP engines. This is because the former agent now has an Access Control Subsystem added to protect unauthorized access to its MIB.

The four subsystems are:

- Message Processing Subsystem – extracts data from messages;
- Security Subsystem – security processing on messages, concerned with privacy and authentication;
- Access Control Subsystem –concerned with authorized access to specific MIB objects;
- Dispatcher – prepares messages for transmission.

2.3.4.2 SNMP Applications

The five applications for generating and receiving messages are as follows:

- command generator – prepares a request after security checks and matches the response against the request, carrying out any subsequent action required;
- command responder – prepares a response to a request to carry out an action after checking the sender's credentials and carries out that management action;
- notification generator – handles *InformRequests* to other SNMP Managers and *traps*, acting in the same way as the Command generator;
- notification receiver – registers to accept and respond to *InformRequests* from other SNMP Managers and accepts *traps*, acting in the same way as the Command responder;

  - proxy forwarder – forwards four types of SNMP messages:
  - forwards *request* to target SNMP engine or to one that is closer to the target;
  - forwards *notifications* to those engines that should receive them;
  - forwards *responses* to previously matched *requests*;
  - forwards *reports* to the initiator of a previously matched *request* or *notification*.

Note: *report* is a new SNMPv3 engine-to-engine communication, which is generated by the Message Processing Subsystem if errors are detected on incoming messages. These can then be forwarded in a preconfigured way for processing.

2.3.4.3 SNMP Security Models

The difference between the two models is the level at which they operate. USM operates at the message level, whereas VACM operates at the lower data level where management information is stored in the MIB. A full description of both models is well documented by Stallings [9].

The security concerns voiced about SNMP from its inception have taken many years to resolve. The current model addresses some threats identified at the time, but as methods become increasingly more sophisticated and appear in quick succession, it is probably inevitable that agreed standards to protect against security attacks will always be one or two steps behind.

The model does, however, guard against modification of in-transit information, masquerading attacks from devices pretending to be a real device on the network, replayed or duplicated message streams and observing values or commands pertaining to password setting.

Denial of service attacks and traffic analysis are not catered for, as it is thought that a more robust firewall technology exists to prevent this.

With the advent of business models, which rely on inter-company traffic between disparate networks [such as is the case with E-business or business-to-business (B2B) transactions], the need to manage such transactions does not evaporate beyond the firewall. Equally, the monitoring of applications' response times must also now be measured past the firewall across the Internet. This posed one of the largest challenges in finding an all-encompassing solution to suit all scenarios regardless of topology, medium, or transport route.

The main features of the security models are as follows. USM provides:

- privacy through encryption;
- authentication technique;
- management of authentication and privacy keys;
- time function – to ensure that responses are received within a reasonable time frame;
- definition of a message format to cope with all of the above;
- discovery procedures between SNMP engines.

VACM effectively splits out the functions necessary to perform configuration changes, for example, into manageable pieces. It identifies and authenticates against:

- which SNMP engine is requesting the action;
- which engine will carry out the action, by itself or on behalf of another engine;
- where in the MIB the information to be changed is located;
- authentication information;
- permissions to carry out the action (Access Control) on that part of the MIB;
- from what level an engine is allowed to view part of the MIB tree[4].

None of these security facilities deliver real-time application response time monitoring, but they do provide a secure framework for monitoring outside firewalls where security considerations are paramount.

Until SNMPv3 becomes more widely acceptable to vendors, the reliance upon SNMPv2 as a working protocol is now assumed for the rest of this document.

2.4 Management Information Base (MIB-I, MIB-II and Extensions)

This section will introduce the SMI directory tree but will focus only on those parts of the MIB that relate to applications-level monitoring. As MIB-II has superseded MIB-I successfully by simply adding more groups to the existing standard, I will refer only to MIB-II throughout.

The structure of the MIB gives us our first hint that applications monitoring has been at least considered in the design, although only IP applications come under its umbrella.

It will also be seen that, while being able to categorize applications by port or socket

[4] This facility has been available as part of a Unix implementation of SNMP but is infrequently used.

number, MIB objects remain a slave to their underlying SNMP protocol, which, it will be remembered, only treats the concept of time very 'loosely' (see Section 2.2.4). This makes tracking the timing of application response times too arbitrary and does not give due consideration to the fact that applications often rely on millisecond response times and that humans increasingly have come to expect as much.

The MIB is also limited to the device where the SNMP agent is resident. In order to monitor an entire network, every device would need to have an SNMP agent installed and, in some cases, report back on behalf of (or *proxy* for) another device that did not support SNMP.

Exactly what constitutes 'a client application' needs clarification before proceeding further, and this is addressed immediately below.

2.4.1 Client Applications – A Definition

For the purposes of this document, a client application is a set of interactions between a client (PC or Unix workstation) and one or more servers using TCP as the transport mechanism for guaranteeing delivery of packets safely to their destination and any one of the IP protocol suite of applications (FTP, Telnet, SNMP) or a custom application that is written to run over IP. That custom application, like SNMP, may employ UDP as its transport mechanism[5]. Each interaction is made up of individual transactions.

It should be noted at this stage that a single key press on a keyboard does *not* amount to one transaction. It frequently can generate scores and sometimes hundreds of transactions between client and server or servers. Each return journey made between client and server is sometimes referred to as a 'network turn'.

2.4.2 SMI Directory Tree

This section describes the four main groups of the directory tree and locates the MIB-II and private MIB branches.

Private MIBs are mentioned in passing because of their contribution to network management generally and their necessity for full functionality of a NMS. It is a testament to the success and flexibility of the SMI structure that so many supplementary MIBs have been developed in both *experimental* and *private* areas. No NMS solution can be said to be complete without the *private* MIB component, but this has also had its drawbacks.

In the directory tree under the *Internet* group are to be found four others:

- *directory* – reserved for OSI X.500 directories;
- *mgt* – Internet Activities Board (IAB) approved objects;
- *experimental* – testing bed for new MIBs before being promoted to the *mgt* group;
- *private* – vendor-specific, differentiating, added-value area.

The standardized set for MIB-II resides under *mgt*, including the *rmon (mib 2)* group discussed in Section 2.5.

[5] UDP does not guarantee delivery of packets. It cannot be excluded from monitoring, as many applications use a combination of UDP and TCP. This poses significant challenges when trying to track application response times.

2.4.2.1 Private MIBs

This area of the SMI directory tree allows vendors to write MIB definitions for their own equipment. This has spawned many useful features that include monitoring of power and environmental management, configuration and management of VLANs, and statistical analysis.

The unfortunate by-product of such talent being let loose in the open house that the SMI structure afforded was a proliferation of network management products tailored to manage just a single vendor's equipment, which then rapidly deteriorated into management of individual types of device as new technologies were promoted: VLANs, Layer2, and Layer3 switches, to name but three.

Adding private MIBs to network devices or 'instrumentation', as it is known, is a time-consuming process and was not always the first thing to spring to mind before products shipped in the 1990s. Such were the advances in technology during that period that it was not uncommon for multiple teams to be formed to work on separate technologies that all came up with their own management solution.

Fortunately, the millennium has begun with a rationalization of tools and the realization that a network management framework is the key to the successful running of medium-to-large enterprises. (See Chapter 6: An Integration Model)

In addition, and very helpfully, some vendors are now producing MIBs specifically with applications monitoring in mind.

2.4.2.2 MIB-II Location

Underneath the *mgt.mib-2* branch are 10 subcategories or groups: system, interfaces, at (address translation), ip, icmp, tcp, udp, egp, transmission and snmp.

This document will now focus on groups 6 and 7, *tcp* and *udp*, because they are of main interest when considering custom applications. Although SNMP can be seen to have a group all to itself, it is transported over UDP and is therefore considered under that heading.

Icmp is also left to one side, because, although ostensibly returning response times from one host to another using PING or extended PING, the way in which these packets are treated by network devices, particularly if small packet sizes of 64 bytes are used, is substantially different to 'normal' applications traffic. Effectively, one cannot rely on PING as a means of verifying application response time. Figure A.1 in Appendix A clearly demonstrates this.

2.4.3 MIB-II Groups

This section gives a brief overview of what kind of data are allowed to be passed up from the MIB and identifies the two subgroups of interest to us. This leads to an important step forward in the identification of an application.

One of the limitations of MIB-II standard is that it relates to individual devices only and does not permit a view of a LAN segment as a whole. This was remedied with the addition of the RMON MIB (see Section 2.5).

2.4.3.1 MIB-II Capabilities

As seen in Section 2.3.3.1, SMI provides the syntax to work with and explains to both how each MIB object is defined and its place in the directory tree. This was to keep SNMP the 'simple' protocol it set out to be.

Apart from defining data types and values or ranges, SMI also lists the capabilities of exactly what can be extracted from the MIB. These are limited to returning either an *ipAddress*, a *counter*(32 bit[6]), which increments until it is full and then starts from zero, a *gauge* (32 bit), which can go up and down like a speedometer but will stick at the maximum once reached, and an *opaque* data stream for arbitrary data or a *timetick*, which relates to time (still 'loose') in hundredths of a second.

None of these really help our cause in determining response times and the one feature that *gauges* could have been used for, namely *thresholds*, was not included in the MIB-II definition, as it was thought NMSs would be flooded with *traps*.

We will return to the importance of thresholds under Section 2.5 when they were included in the RMON MIB definition.

2.4.3.2 MIB Groups *tcp* and *udp*

This section identifies the usefulness of the two groups and also the challenges that it throws out in the search for adequate application monitoring.

Those with MIB browsers open at this juncture can point them at the following part of the MIB tree of a network device to see the two groups:

- 1.3.6.1.2.1.6.13.1 iso.org.dod.internet.mgt.mib-2.tcp.tcpConnTable.tcpConnEntry;
- 1.3.6.1.2.1.7.5.1 iso.org.dod.internet.mgt.mib-2.udp.udpTable.udpEntry.

The relevant entries for monitoring applications are:

- TcpConnLocalPort – local port number;
- TcpConnRemPort – remote port number;
- UdpLocalPort – local UDP port number.

With this information, one can monitor certain characteristics of an application, such as who is talking to whom later exploited in the RMON MIB. This is because, in order for a client to communicate with a server (or servers) using TCP, a negotiation first has to occur between them. Amongst other things (such as packet type, size, window), port numbers are agreed as a means of opening communications between the two parties. The combination of IP address followed by an identifiable port number is referred to as a *socket number*.

Although a single socket number can support multiple connections, there can be only one pair of sockets communicating between each other at any given time. This unique quality of TCP opens an important door for monitoring specific applications.

2.4.4 Application Identification

Developers of bona fide applications register a known port number or range of port numbers

[6] This simply means that the number can increase to $2^{32} - 1$ before returning to zero. In practice, troubleshooting often involves resetting such counters to see if the errors, for example, are still active.

for their application, so that no other application can conflict with theirs. These port numbers are normally applied at the server end of a transaction with the client supplying a random port number for the duration of the interaction. Some client-server applications, however, do allocate a static port number at the client end.

This represents an important step forward, because we can now identify an application by a known port or range of ports.

Regrettably, there are exceptions to this behaviour, but the vast majority adheres to the registration convention or gives the implementer the option of allocating static ports. This is a necessary feature of any application that wishes to pass through a firewall.

In conclusion, then, MIB-II on its own is not enough. It can identify certain applications by port and socket number on a single device and tell us the destination, but no more.

2.5 RMON and RMON2

This section, like the previous section, will confine itself to those parts of the RMON2 MIB relating to applications level monitoring. Apart from stating the categories in the initial RMON specification, its successor, RMON2, will be the focus of this section. This is possible also because RMON2 was simply an extension of the same *rmon (mib-2)* group.

As stated in Section 2.4, MIBs alone could not monitor large networks, as a NMS (or multiple NMSs) would have to request large quantities of information on a regular basis and present it, as well as processing any *traps* that were sent to it in order to serve any useful purpose.

Enter Remote Network Monitoring (RMON), a means to monitor whole LAN segments in one go and have data stored locally and supplied to an NMS on demand. When coupled with new *alarm* and *event* capabilities, an RMON *probe,* as they are now referred to, could send a *trap* to an NMS accompanied by the relevant traffic data warning of errors, high utilization, or congestion.

However, one limitation remained with both RMON and MIB-II: an inability to measure the impact of what is sometimes called 'tandem traffic', that is, traffic passing through a LAN segment originating and terminating on unmonitored remote segments of a network. This is because RMON operates only at layer 2 of the OSI model, the physical or Media Access Control (MAC) level, and therefore can report nothing about network layer traffic and above.

For that, the RMON2 standard was developed.

2.5.1 RMON2 MIB Location

In the SMI directory tree underneath the *mgt.mib-2.rmon (mib2)* branch are 19 subcategories or groups, of which the first 10 historically belonged to the initial RMON specification:

RMON:

- *statistics* – utilization and errors;
- *history* – provides statistics at user-definable intervals[7];
- *alarm* – allows the setting of rising and falling thresholds at sample intervals;

---

[7] RMON2 added the network efficiency of providing only the changes in historical statistics for the next time slot, rather than resending the same counters again.

- *host* – keeps a log of hosts (layer 2 MAC addresses) and counts packets, octets, errors, broadcasts, and multicasts in and out of each host;
- *hostTopN* – sorts the *host* statistics to give 'top talkers' on a LAN segment;
- *matrix* – puts *statistics* into a matrix to show who is talking to whom;
- *filter* – allows the filtering of packets captured or associated statistics;
- *capture* – defines what and how many packets are captured;
- *event* – stores events issued by the probe and may cause a *trap* to be sent;
- *tokenRing* – statistics and configuration information relating to a single ring on a Token Ring network.

2.5.2 RMON2 MIB Groups

RMON2:

- *protocolDir* – catalogue of all protocols the probe knows about;
- *protocolDist* – aggregate statistics per protocol per LAN segment;
- *addressMap* – maps network addresses to MAC addresses (*host* above);
- *nlHost and nlMatrix* – network layer host statistics and pairing combinations;
- *alHost and alMatrix* – host statistics and pairing combinations at transport, session, presentation and application layers;
- *userHistory* – samples and logs data in user-definable ways;
- *probeConfig* – configuration parameters for RMON probes.

From this, we can see that applications monitoring has arrived.

It must be noted that because of the nature of IP applications, which do not adhere strictly to the OSI 7 layer reference model, *alHost* can refer to anything running above the network layer.

2.5.3 RMON2 MIB Capabilities

This is a vast improvement and provides us with *counters* and *gauges* as before, but this time at application level, with additional capabilities of setting rising and falling thresholds triggering off a *trap* if required.

We can also see who is talking to whom across LAN segments bi-directionally and who the 'bandwidth hogs' or troublesome devices are on those segments. Some vendors have managed to adapt or extend RMON to cover WAN protocols too, releasing RMON2 from its LAN-locked history. To help our cause, all this happens in real time too.

So, does RMON2 fulfil the need to monitor applications' response times in real-time and lead us to the root cause? It certainly is suited to troubleshooting down at packet level and provides utilization of which applications are heavily used and who is using them, but there is *no response time element* that the end user can directly relate to.

2.5.4 RMON2 Limitations

Quite aside from the lack of a response time measurement, there are ever-present problems with RMON probes. Those are questions of speed and ability to cope under stressful network conditions.

As transmission's speeds double and redouble on both the LAN and the WAN, capturing data at high speed and with such fine granularity sometimes overwhelms the probes, and they simply lock up or report incorrectly on what they have managed to gather. It is essential that they remain in service at those times, as that is when one needs them most.

It would not be surprising if, for reasons of speed alone, probes were to remain at the edges of the network, but they have also laid down an important precedent that should not be forgotten: that of monitoring, at application level, from one end of a transaction to the other.

It was this that led me to consider the endpoints of those conversation pairs as a possible source of relevant data for monitoring response times: the humble PC. That workhorse on which we all depend to carry out the daily work tasks became the natural place to consider response times from a customer perspective.

Before progressing to look at current methods of addressing this, CMIS should be discussed.

2.6 Common Management Information Services (CMIS)

This section outlines how network management is addressed in the Open Systems Interconnection (OSI) environment. It introduces the elements of the framework designed to monitor and manage an OSI network, based on ISO protocols.

2.6.1 Summary

It will be seen that it is not too dissimilar in purpose from the SNMP view, and some elements of the SMI structure are shared. Their respective implementations diverge quickly afterwards, but a working solution that delivers a framework capable of handling large networks, largely in the collection of relevant and correlated data to enable root causes to be arrived at swiftly, escapes both camps.

Furthermore, it is not evident that application response time is among the retrievable data.

OSI-based environments have their own standard Common Information Management Service/Common Management Information Protocol (CMIS/CMIP). In an attempt to quantify and cure all network ills, this is disadvantaged by its sheer complexity and is not therefore favoured by most vendors.

2.6.2 Attempts to Combine CMIP and SNMP

Attempts were made to combine the CMIP over TCP/IP (CMOT) and SNMP approaches, but these had to be given up, as the two are different in design: CMIS and the protocol that carries it (CMIP) are moulded around an object-orientated database structure. The way in which SNMP uses SMI is essentially a static database, a series of arrays with associated attributes.

The condition that there be a common SMI/MIB was dropped by the IAB [8], and the two went their separate ways.

2.6.3 OSI Management Model Overview

The OSI management model [10] has similar functions to those used in the set of standards under SNMP. We need to remember two things:

- We are dealing strictly with the seven-layer OSI model, including transport, session, presentation, and application layers;
- We are in the realms of dynamic, object-oriented databases with Managed Objects (MOs) and pointers between them

There is a heavy reliance on agents. The SNMP agent has become the Local Management Application Entity (LMAE), but its role is far greater in that the agent is able to dynamically create, delete, retrieve, or change MO in the object-oriented database, which is known as the Management Information Tree (MIT). This is based on the X.500 directory tree [11] and is not unlike the SMI directory tree. MIT trees have parents and siblings.

A unique distinguishing name (DN), a concatenated version of its relative distinguishing name (RDN), is used by CMIP to identify a device and to access management information. The RDN could be a port identifier on a hub to distinguish it from other ports on the same hub.

Agents have the ability to interrogate individual layers of the seven-layer OSI model. However, the routes and procedures that have to be invoked are not simple. Figure 2.5 illustrates the CMIP communications flow and some of the MIT architecture.

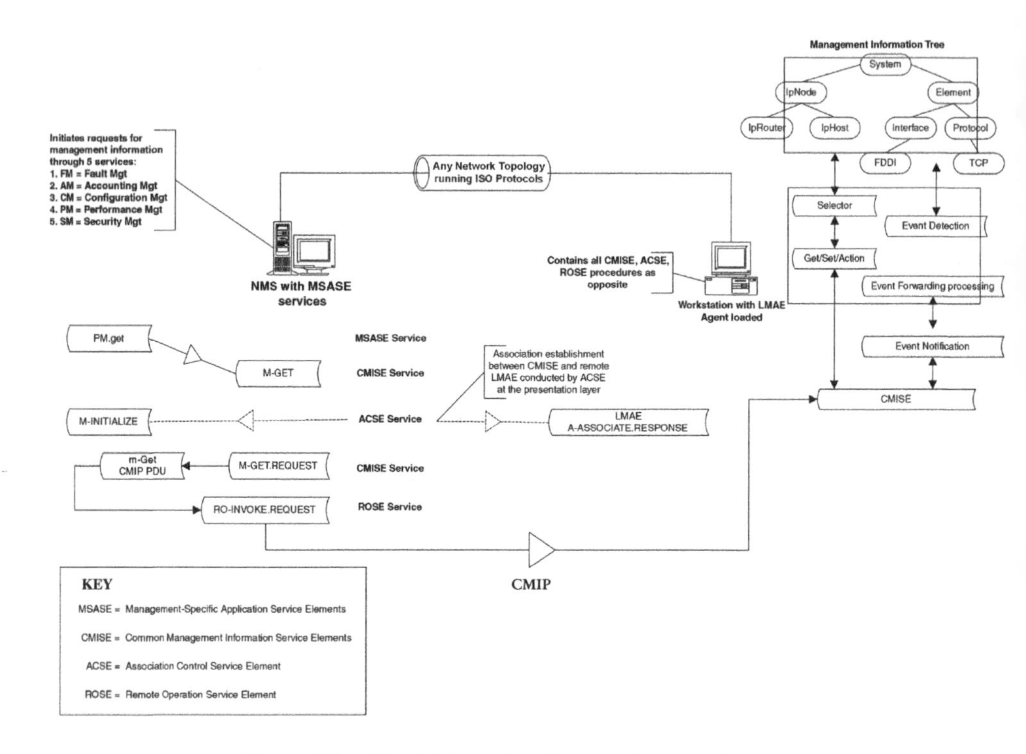

Figure 2.5 CMIP information flows and MIT architecture.

What amounts to a single *getRequest* in SNMP has to go through six different iterations between the various service elements, each with its own procedure and syntax, before being passed to CMIP for transporting to its destination.

It is little wonder that the reasons for the complexity were lost on the majority of vendors.

2.6.4 Considerations and Concerns

There is a strange paradox in this model in that it uses connection-oriented transport protocols that guarantee delivery, normally considered to be a good thing and certainly an improvement, one might think, on the 'unreliable' UDP protocol used to transport SNMP. However, what is one man's strength is another's weakness, for the additional processing this particular protocol imposes on the device proves to be its undoing.

The flexibility and efficiency of an object-oriented database that dynamically links different parts are lost in the complexities of its implementation and the raw processing powers required to store and manage this information. Also, concerns over data corruption brought about through the deletion of some objects, which still have pointers to other objects, are well justified.

In addition, of the two added advantages that CMIS had over SNMP of bulk and selective retrieval of management data, the former was removed with the *getBulkRequest* of SNMPv2 and the lack of perceived added value of the latter consigned CMIS to the role of spectator in the network management arena.

One commentator assesses CMIS in this way:

> OSI provides an extended OO (Object Oriented) database framework .... Management information bases, however, need to balance conflicting requirements for functionality and real-time performance under resource constraints. It is not yet established whether the OSI design choices can strike such balance [12]

So, real-time performance remains at the heart of what every NMS should deliver. Without timely warnings of performance degradation, proactive management will remain the stuff of dreams.

2.6.5 CMIS Capabilities

This model offers a different method for looking at similar data to SNMP in that, in spite of all its complexities, it still only returns counters and gauges. Hayes points out the limitations of the CMISE model in relation to the need to gather data over different time intervals [13]. The object-oriented design is not good at handling data held in arrays, which the collection of historical data requires. Hayes concludes that extracting data relating to a specific time period is difficult, and accessing individual data within arrays impossible.

Once the connection between NMS and LMAE has been set up, it is used for the following purposes:

- Event Reporting – LMAE initiating event data transfer to NMS;
- Information transfer – NMS requesting statistics from an LMAE;
- Control – NMS sends selected actions down to be carried out by the LMAE.

Unfortunately, this takes us no further forward in terms of monitoring application response times in real-time.

It is time to turn our attention in Chapter 3 to current methods including Application Response Monitoring (ARM) and then move on in Chapter 4 to look at the latest developments: a project called NIMI and Intelligent Agents.

3

Current Standards

This chapter deals with various standards, current or developing, to establish the continuing need for a different approach to monitoring applications' response time.

We will take a brief look at each of the following in turn and see why they too fulfil neither the goal of application response time management nor the fundamental business requirements of today:

- Systems Management System (SMS from Microsoft)
- Applications Response Measurement (ARM)
- Common Information Model (CIM from the dMTf).

With the proliferation of standards to choose from, it is easy to lose sight of the business needs, which have evolved away from the OSI network management model. So, this chapter ends by looking at the business drivers for such a change in direction and signals why SLAs have raised the profile of response time management.

3.1 Systems Management System (SMS)

This section looks at the limited scope of SMS in the field of applications management today.

Not to be confused with the other SMS (Short Messaging System – a facility found on mobile phones limited to 256 characters), this began life as a system to retrieve information about desktop PCs for inventory and support purposes, which normally necessitated a desk visit. Princes amongst these include such delicacies as versions of the Basic Input/Output System (BIOS – the means for your computer's processor to talk to an external hardware port) and Network Interface Card (NIC) details and driver versions.

Microsoft[1] has added software distribution, data-filled asset management and packet capture to its capabilities. Despite the latter, there are no facilities in SMSv1.0 to monitor applications' response time real-time from the PC.

SMS concerns itself entirely with the management of Windows® systems, and performance monitoring is restricted to servers, anticipating service degradation and causing an event to be generated. That said, SMSv2.0 has also been designed to collect and manipulate data in the CIM format (see Section 3.3), which eliminates the need for data collection agents on systems running CIM-compliant instrumentation, such as Windows 2000®.

[1] Microsoft™ is a trademark of Microsoft Corporation.

In collaboration with a third party, an engine was added to SMSv2.0, which would allow the PC to become potentially a network probe for its own segment and therefore for its own traffic, but no application exists today to make use of that.[2]

3.2 Applications Response Measurement (ARM)

This section investigates why ARM has not captured the imagination of application developers and the business community at large, in spite of its sound objectives.

Possibly the best chance of introducing a standard for applications' response time management was the introduction of ARM, which required developers of applications to include an element of monitoring application responses and general health during a transaction or series of transactions. This is known as *application instrumentation*. Applications that support ARM typically report errors to a console screen or write them to a log. Some NMSs would be able to interrogate the logs and make sense of the information they receive.

Establishing whether an application process has hung on a server has always eluded most systems management systems dealing with the Unix environment. This gave an opportunity not only to track a transaction through each of its *network turns*, but also to define transactions in terms of business processes (See Chapter 4 Latest Developments Section 4.3.3 for a fuller explanation of a *network turn*).

This would certainly have delivered real-time response time analysis of application transactions but would not have been extended to help find the root cause or potential suspects in the case of excessive latency.

3.2.1 Objections to ARM

The main objection to ARM is its inherent invasiveness into the application. I have seen it described as 'open-heart surgery' [14].

Similar concerns were, and still are occasionally, raised over the intrusive nature of network or applications management software and its potential impact on other 'more important' traffic. Much was done to increase the efficiency of data collection and reporting back to NMSs as we have seen. From experience, I would endorse the view that management traffic will normally only utilize 1–3% of network capacity [15]. However, subjecting applications to potentially similar overheads proved too much for most to swallow, and ARM seems only to have been supported by a small number of vendors of management systems.

With the emergence of heavyweight Enterprise Resource Planning (ERP) applications, it has been much more viable and desirable to 'ARM-enable' the application, since these can cross several business unit domains within an organization spanning several inter-related transactions. Even these ARM-enabled applications face an uncertain future as electronic commerce (e-commerce) takes hold, as these typically span several businesses and often involve human, manual intervention lasting days.

[2] The recent merger of Compuware™ and Optimal Networks™ may present an opportunity to change that.

3.3 Common Information Model (CIM)

This section investigates the CIM model, as proposed by the Desktop Management Task Force (dMTf) [16], and outlines the confinements of it with respect to applications management.

We will look at the two parts of the model:

- specification
- schema

and understand which parts of the schema relate to applications management and why these are not appropriate for identifying response time degradation or for determining root causes by themselves.

3.3.1 CIM Model

This model is defined, first, as being a conceptual model for management that is not bound to a particular implementation and which is therefore a standard.

Second, it aims to pass management information between agents and managers or between managers and managers to provide distributed systems management, in much the same way that previous standards considered above have striven to do in the past.

It combines the strengths of the MIB extensions success with the flexibility of an object-oriented format, as in the case of CMIS. The goal was to permit the development of applications that would recognize different vendors' equipment and base software from management data provided by each of those vendors in the same CIM format.

3.3.2 CIM Specification

This describes the naming conventions, language and mapping techniques to exchange information with SNMP MIBs, CMIS MITs, and now dMTf MOFs (Managed Object Files). It also included a Meta Schema to define terms used in the model, their usage and syntax.

3.3.3 CIM Schema

These are model descriptions, a way to organize the managed data into different categories or layers. There are three layers:

- core schema – applicable to all areas of management;
- common schema – applicable to particular areas of management but independent of technology or implementation. There are five common schemas:
  - systems
  - applications
  - networks (LAN)
  - devices
  - physical

- extension schemas – technology-specific areas fall into two categories:
  - specific general – authorized by dMTf
  - product specific – registered with the dMTf.

The formal definition of each schema is held as a MOF, which is an ASCII file. MOFs come with their own MOF editors, parsers, and compilers to be used in other applications.

A product is said to be 'CIM-compliant' if it implements the core and appropriate common schema as a minimum.

3.3.4 CIM Capabilities in the Future

CIM's stated goal is to unify existing standards using an object-oriented structure. It would also like to use the subclasses inheritance of information from classes above (as with other parent–child relationship structures) to pass on properties and associations with other objects in order to solve problems across traditional vertical domains of system, device, network, user, and application.

In addition, it aims to deliver the management information in a web-based format: Web-based Enterprise Management (WBEM).

3.3.5 CIM Capabilities Today

One can see CIM-based management products today, and they are from the major vendors who go to make up the dMTf. The core model is well defined down to the symbols and colours used to describe the 'is–a, has–a, uses–a' relationships. An example is given in Appendix B.

The Application Management Model is disappointing. It promises to "describe a set of details that is commonly required to manage software products and applications". [17]

Unfortunately, response-time management does not feature anywhere in the model below. Software is defined as having four components:

- Software Product – collection of software features licensed to buyer;
- Software Feature – selectable components to perform a function made up of software elements;
- Software Element – files associated with the software;
- Application System – collection of features that combine to support a business function.

It defines how much memory is needed to run an application, disk and swap space requirements, OS version, files, and directories, all that goes to make up a now common installation assistant that comes with software. Other features are whether an application is *installable*, *executable*, or *runnable*. These one can see as useful tools from an applications service provider's (ASP) perspective when wishing to control the downloading and licensing of software to remote machines.

In brief, this has more to do with software distribution, inventory, and metering (tracking the legal licensing of software) than response-time measurements. It has little, if anything, to offer today in that space, but is gathering momentum not least because there is much to be gained from putting definitions and schema in place, which other systems can query successfully every time without employing skilled individuals to extract the data. In a fully ordered universe, it would be easy to see the benefits of directory services that feed on such schema, as is proposed in some of the emerging policy service models.

We saw before how one business-process-orientated standard, ARM, was struggling to find favour. It is curious how so many systems or frameworks pay lip service to the fulfilment of business needs, but precious few measure up to that responsibility.

One business factor trying to make its way to the fore is the SLA, which, although not implemented in many organizations, is forcing its way up agendas. The next section looks at why this is.

3.4 SLA Factor

Once the preserve of telecommunications service providers, the SLA is finding its way into ordinary enterprise organizations that find themselves swept along in the stampede to be 'on the web'.

The visibility this affords businesses great and small is a double-edged sword. On the one hand, you have potentially a planet full of customers who could come knocking at your portal; on the other hand, those same customers at best will be the first to complain to *you* about the slowness of *your* site (regardless of the Internet in between) and, at worst, will vote with their feet and never return.

Business units (BU) have reluctantly agreed to internal Information Services (IS) organizations not providing adequate SLAs either because a dearth of tools have been proffered as an excuse, or because the information put before them related to servers, application processes, and network elements. This has nothing to do with the simple requirement to know whether an application is available to all users or whether its response time is below par.

With the advent of firstly outsourcing and then of e-commerce, those same applications are now being offered to external customers (or to the outsourcing company) who demand compensation for lack or degradation in service. Customer service is uppermost, but the financial and commercial penalties of failing to proactively monitor response times and repair faults before they become critical are hard to ignore.

> "As recently as 10 years ago, the majority of IT service contracts did not include service-level agreements (SLAs) traditional measurement schemes are evolving to a new level and are being quantified in new ways. Penalties and incentives are increasingly pointed toward business-based benefits The new penalties and incentives use extent and duration, tied to business impact, to ensure that enterprises are paying for impact, rather than effort expended." [14]Source: Gartner, Inc, May 2001

Source: Gartner, Inc, May 2001

3.4.1 SLA Factors Deemed Most Important (2000)

This has been borne out by a most recent study of 150 medium-to-large enterprises conducted by Infonetics Research, who also rated how external SLAs are currently or will be defined once in place as shown in Figure 3.1.

This clearly demonstrates the validity of finding efficient ways of monitoring response times both for network and applications management. If the two could be combined in one

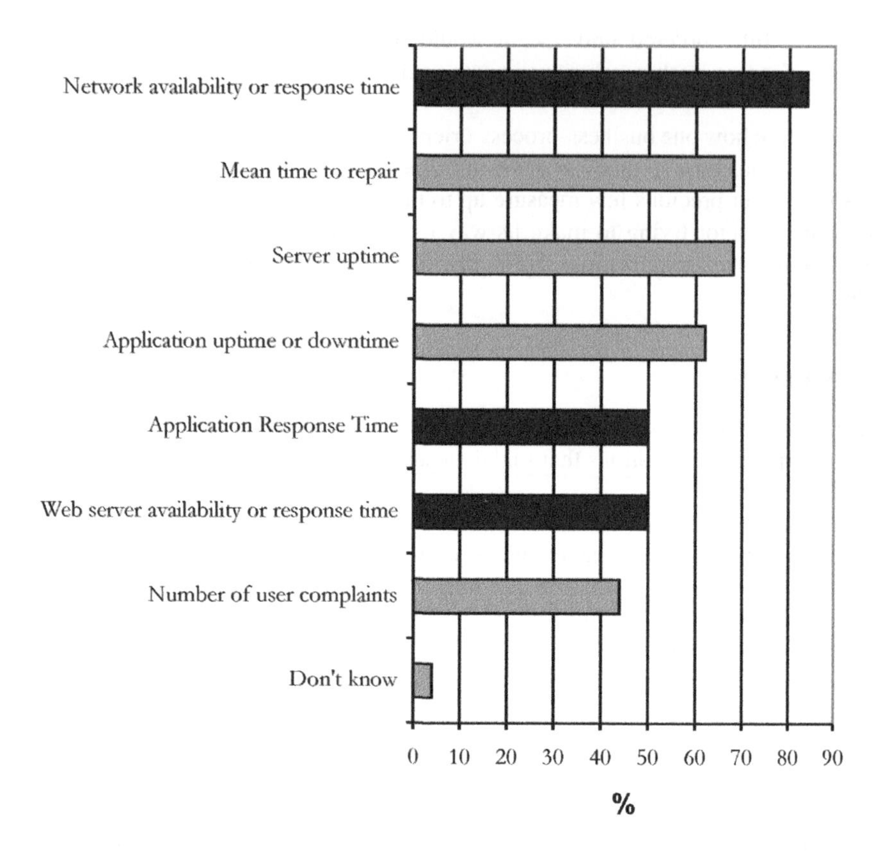

Figure 3.1 Most important SLA factors (2000). Source: Infonetics Research *User Plans for Network Management 2000.*

management tool or fed into a distributed network management framework, a solution would be near. Tying down that degradation of application or network performance to a component or components of the network would provide the root cause analysis we are searching for. The addition of application or server uptime or downtime (or availability in SLA-speak) would be a bonus.

One such tool exists today, which I will focus on in Chapter 6 (An Integration Model) to show how relatively simply these kinds of measurements can be made with little expense and low administrative overheads, and how, with minimal effort, they can be integrated into a network management framework.

4

Latest Developments

We saw in RMON2 Limitations (Section 2.5.4) how probes set a precedent for monitoring end-to-end conversation pairs but also how the sheer volume of data they collect can overwhelm them. While capable of analysing specific problems, once an idea of which end devices are to blame is established, it can hardly be said that probes are proactive by nature.

This chapter looks at the different approaches taken to measuring application response time that have led to truly proactive monitoring and also looks at the rise of intelligent agents, inspired perhaps by an Internet project called NIMI.

The following sections cite the strengths and weaknesses of each of the following approaches:

- Application Instrumentation;
- Network Probes (or 'X-ray' tools);
- Active Monitoring (or Capture and Playback);
- Passive Monitoring (or Client Capture).

The chapter then moves on to the NIMI project, its goals and the lessons learnt, and how one company took this several steps closer to reality.

4.1 Application Instrumentation

This section will not add much more to Section 3.2 (ARM). Suffice to say that the opportunity to define transactions in terms of business processes has largely been submerged in the objections to its inherent invasiveness referred to earlier. It is indisputable that such instrumentation would add to the overhead and impact the runtime performance of an application. The fact that so few developers have produced ARM-enabled applications tends to indicate that ARM does not have such a distinguished future ahead of it.

4.2 Network Probes (or X-ray Tools)

This section sets down their capabilities and strengths as well as the reasons why the probe approach has not been popular in all but the large enterprises where costs are not necessarily a limiting factor.

4.2.1 Capabilities

Their strengths stem largely from their proprietary implementations of remote monitoring (RMON2) set out above in Section 2.5. Unlike ARM, they are not invasive but simply observe traffic as it passes them by.

Probes are divided into LAN and WAN varieties, remembering that RMON was a standard only applicable to a LAN segment originally, but expanded under RMON2 to take in 'tandem' traffic originating and terminating on either side of that LAN segment. Hence, any adaptation for the WAN is proprietary by nature.

They are hardware probes that read the header of packets and match them against known protocols and port numbers in a knowledge database of applications. Minimum, average, and maximum response times are calculated during user-defined intervals. Some operate at higher levels, capturing session, database system management level and even SQL statement-level data, displaying it on a link-by-link basis. This permits us to see heaviest users of a link (*topNHosts*) and also the most heavily utilized links.

4.2.2 Limitations and Constraints

One limitation is the inability to match all traffic. The Other or Unknown segment of a pie chart tends to be the largest on offer, since many applications do not obey the simple rules of port assignment as seen above (Section 2.4.4 Application Identification).

Another is the lack of user perspective in the data captured, and this approach takes no account of delays present in a user's desktop PC.

These probes are usually prohibitively expensive for most small-to-medium enterprises, as they require large numbers to be deployed at strategic locations to capture traffic at the junctions of major data highways.

4.2.2.1 Probe Positioning

The positioning of these probes is of concern too. For those probes wishing to monitor just LAN traffic, life was easy under single broadcast domains on a shared Ethernet segment. Once the idea of devoting one device per segment became commonplace (switched Ethernet segments), this sent the LAN probe manufacturers scrambling to find a means of monitoring multiple switched Ethernet ports on a hub, for example. Their only two answers, first, to copy traffic from the port you wish to monitor to another spare port (or *mirroring*, as it is known) and, second, to monitor the uplink from the hub to a main switch were never going to convince network managers to invest in such a solution.

4.2.2.2 Effect of Ever-increasing LAN/WAN Speeds

The last constraint already alluded to is that, in both cases, the LAN and WAN probes are hard pressed to keep up with today's speeds. Faster interfaces are produced to capture transactions in the fraction of a second it takes to dip into the torrent of data passing by, decode them to layer 7 of the OSI seven-layer model and present them in an intelligible display. This takes memory resources, processing power, and disk space to store the captured data long enough for analysis.

4.2.3 Summary

An industry (Data Warehousing) has grown up around the processing of masses of data thrown out by these probes, but to what end? They may be able to provide trend data for analysis after a night's full processing, but the immediacy of the problem one is trying to solve has long since disappeared.

The probes are good at what they do: capture the fine granularity to define to the last bit and byte what the root cause of a problem is when you have the luxury of it being a persistent problem over time.

They do nothing to warn of degradation in application performance to assist in proactive monitoring and can only cover a large network at considerable expense.

4.3 Active Monitoring (or Capture and Playback)

This section first defines the differences between active and passive monitoring. It then goes on to describe a typical, simple network scenario and its various access methods, before looking at the strengths of capture and playback, what is used, and to what effect in the pursuit of real-time application monitoring.

It will be seen that it is an important component in gathering regular samples of transactions intended to mimic and, in some cases, take the place of real transactions. There are also certain considerations one must take into account when interpreting results from this method.

4.3.1 Differences between Active and Passive Monitoring

Active means that *traffic is placed on to the network* deliberately to measure response times at application level.

Passive means that *traffic is observed*, as in the case of the Network Probes above (Section 4.2), but this time *from a user perspective*.

4.3.2 Simple Network Access Methods

Figure 4.1 illustrates the ways in which applications traverse a network.

This depicts a simple architecture where one desktop PC talks to a single server. In reality, today, web-based applications frequently send out *streams* of transactions simultaneously to different destination servers, and the returning packets then reconvene and re-assemble to form the web page before you.

4.3.3 Strengths of Capture and Playback

In Figure 4.1, the simplest of transactions involving just a single key press results in several exchanges across the network between desktop PC and server. These are sometimes referred to as *network turns*.

There are two ways of monitoring application response time by placing traffic directly on to the network:

- directly querying a server using playback;
- simulating applications traffic without querying the server.

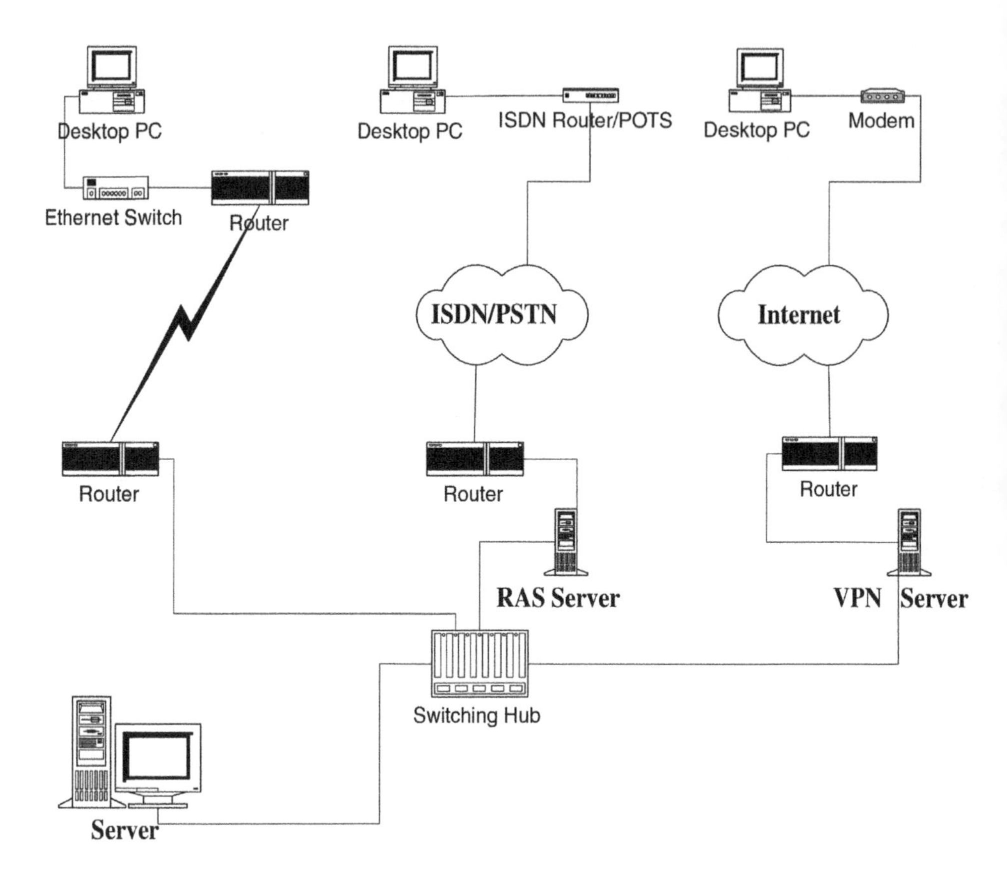

Figure 4.1 Routes an application may use to traverse a network.

The first submits real transactions to production servers and, as such, should be treated with care. It could easily be accused of the same heresy as ARM, namely, adding overhead to an already-busy server, and find itself subjected to a similar fate.

The second transmits traffic between two dedicated endpoints, which do nothing else but pass applications traffic at regular intervals. This traffic mimics a real transaction or sequence of transactions and is often known as a *synthetic transaction.*

Regular transactions mean consistent data collection suitable for anything from trend analysis for application or network planning purposes to verification of SLAs.

The advantage of the second approach is that it effectively eliminates both desktop PC and server from the equation. This then gives a true picture of the effect the network portion is having on applications traffic, measuring response time as it goes. This cannot, however, be considered as 'real-time' in the way the user experiences it, because the regular measurement and the user experience would not normally coincide.

It will be seen later how agent technology can deliver so much more than response time information, which ultimately will aid root cause analysis.

4.3.4 Active Considerations

It must be remembered from Figure 4.1 that the 'network' in the case *synthetic transactions-* would include both WAN and LAN segments situated between the two endpoints. Hence, any local LAN problems would potentially cloud the issue.

It is also important to realize that your choice of endpoint platform is important. A PC with sufficient memory and processing power, albeit minimal, would return dubious results if it were being used to play games over the lunch period!

4.4. Passive Monitoring (or Client Capture)

This section describes the different perspective that client capture brings without the need for intrusive application instrumentation. We look at the relative strengths when compared with the active approach and considerations for the different type of data produced, to understand its place in the monitoring of real-time application response time.

When it comes to getting close to the customer and what pain they are feeling through performance degradation, it does not come much better than this.

4.4.1 Strengths of Client Capture

As client capture is carried out at the desktop PC, the strengths are principally that the data being collected is extremely detailed, can be tied down to the second slowdowns occurred, and provide valuable user workflow data.

Some vendors concentrate on web response times, monitoring page retrieval times or downloads, and using simple web page queries, whereas others monitor any application with a known port or range of ports by means of intelligent agents. The usual caveats apply about 'rogue' applications.

It is important to understand that the type of data now being collected is sporadic in nature. An application used once a day will record only one point on a graph each day. These data then are not suitable for trend analysis and no substitute for regular synthetic transaction testing. Rather, the two complement each other.

4.4.2 Passive Considerations

Given that we are now looking to monitor desktop PCs, several questions arise. How many? Where? How can one accommodate a mobile workforce and many more besides?

For now, it must be said that putting agents on all desktops in a large company would be folly, resulting in torrents of data again.

It is also perhaps wise to remember that users have different working habits, may have different PC configurations or have different combinations of applications running at any one time, when interpreting the data. Also, users of the same application may fulfil different tasks, one resolving problems involving multiple small lookups from a database and another possibly transferring large amounts of data regularly.

What one can say is that when policy-based Quality of Service (QoS) is applied on a per-user basis, this approach is well placed to monitor an application from the end-user perspective. This is largely owing to the agent technology used.

The next section looks at a project that typified this approach by using *daemons* (processes running under Unix). Its name was NIMI, or the National Internet Measurement Infrastructure.

4.5 NIMI Project

To date, we have really confined ourselves to applications running across LANs and WANs (predominantly using private leased lines), but degradation in application response time across the Internet has become more of a driver to find a solution that fits all networks irrespective of their components.

This section describes the project overview, the goals of NIMI, how the architecture proposed showed similar characteristics to the agent software now beginning to emerge on to the market, and the lessons that were learnt.

It will be seen that it is an important step towards making real-time application performance monitoring of large networks a reality. The particular network this group had in mind was the Internet!

4.5.1 NIMI Project Overview

Born out of a need to monitor traffic performance across the Internet, NIMI stated its goals as being to facilitate

> ... a 'measurement infrastructure' for the Internet, in which a collection of 'probes' co-operatively measure the properties of Internet paths and clouds by exchanging test traffic amongst themselves... [18]

The main aim was to tackle the issue of scalability, still an unresolved issue with many traditional network management systems (NMS) today. Principal among the measurements they wished to make were connectivity, throughput, latency and packet loss rates. These all survive today in the agent software cited in the Chapter 5 An Integration Model.

The key to providing timely data about network health could not be constrained by the limits of how many times a manager could poll managed entities or by how much data a managed entity could reasonably hold. To this end, the concept of managerial control had to be exploded.

4.5.2 NIMI Architecture

With the key emphasis on scalability, the architecture [18] reflected the desire for:

- decentralized control of the measurements;
- strong authentication and security to control who ran what tests from which probes;
- configuration and maintenance of probes;
- delegation of measurements outside of a site's normal boundaries.

The second and third of these come as little surprise, and even the last is reasonable to expect when dealing with measurements across different Internet Service Provider's (ISP), traditional telecommunications Service Provider's (SP), and company's domains.

The first of these was essential to overcome scalability issues.

The project used NIMI daemons (or *nimids*) to carry out tests remotely. The focus on end-to-end measurements meant that entire paths across the Internet could be managed without regard for who owned any portion of it. The *nimids* were required to be:

- 'high-yield' in retrieving maximum data from the smallest of tests, meaning that *active* measurements would impact performance least;
- lightweight and portable to enable any platform to carry out the tests;
- dynamic in carrying out changes to the tests required in a seamless fashion;
- secure;
- able to self-configure.

All these characteristics are present in the agent software.

They also share similar coded shorthand to enable them to carry out scheduled tests in the most efficient ways. In the case of TCP, streams of transactions involving multiple *network turns* contain 'control traffic' to guarantee safe delivery of each packet or sequence of packets and to ensure that they arrive in the correct order. The acknowledgements (ACKs) that make up a large part of this 'control traffic' are very small packets, and because they can acknowledge groups of packets rather than every packet, they do not add any delay to other packets and can be safely ignored when constructing *synthetic transactions*.

4.5.3 Lessons Learnt from NIMI

The notion of the importance of time raised its head again and needs now to be considered. One remembers the vagaries of 'SNMP time', but now, to accurately record within milliseconds an application's response time, real or *synthetic*, requires that both machines at either end have their time synchronized.

The group conducting this in 1998 did not have the luxury of deploying Global Positioning Systems (GPS) because of the expense and system installation difficulties, but GPS systems are much more reasonable in price today, and with the assistance of the Network Time Protocol (NTP), one can easily distribute time at Stratum 4 level (Stratum 1 being the atomic clock level) without too much difficulty.

Exhausting system resources was a major problem, because the programming they did in C or C + + was prone to memory leaks. Java was thought to be a 'safer' language.

Domain Name Systems (DNS) also created problems when the DNS servers were unavailable for resolving hostnames. This has become increasingly important as Dynamic Host Configuration Protocol (DHCP – a system for allocating IP addresses from a pool, obviating the need for some systems to have a static IP address) is now widely used and is wholly reliant upon hostnames.

In order to track application response time real-time from a telecommuter or mobile user, the agent must be traceable wherever it attaches to the network.

One of the drawbacks of monitoring end-to-end with probes is their positioning (Section 4.3.4 Active Considerations). Figure 4.2 illustrates this.

Endpoint A has been set up to conduct regular tests with Endpoint B using a *synthetic transaction*. Without the 'control' test between Endpoint A and Endpoint C, any variations in results between A and B could not be categorically put down to degradation over the WAN. Control tests between A and C would identify immediately whether there is a local LAN problem compounding response times over the WAN.

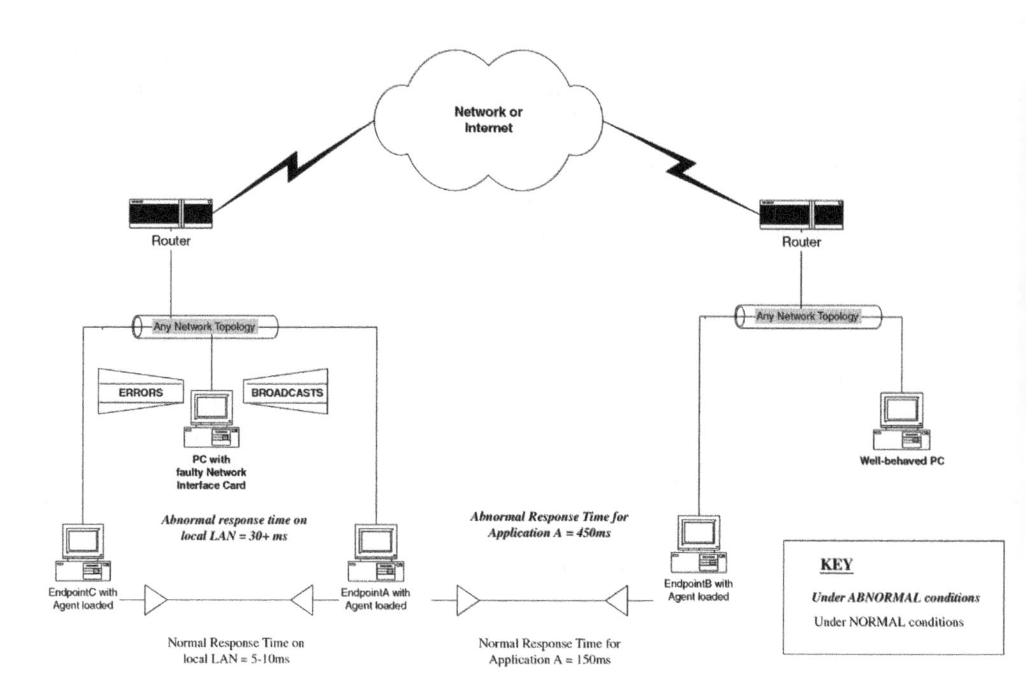

Figure 4.2 Effects of excessive LAN traffic on end-to-end measurements.

The NIMI approach laid the foundation for a scalable solution to monitor response times that involved lean agent technology. We will now see how one such agent technology has been used to provide real-time monitoring of application response times and how it has enabled root cause analysis.

5

A Model Agent

Application response time, by itself, is not enough to pin down the root cause of the degradation being measured. That requires knowledge of how TCP flows are made up to derive maximum information from the least amount of data collection, one of the stated requirements of the NIMI *nimids* (Section 4.5.2 above).

This chapter shows how agent technology has matured into something that measures a good deal more than just application response time, while fulfilling most of the NIMI architectural considerations: a model agent indeed.

We begin with an explanation of the agent architecture and how closely it matches NIMI's requirements. Then, we consider the TCP flows in detail to see how root cause analysis can be derived from the way the protocol works.

5.1 Agent Architecture

This section describes what an agent is, how closely this matches NIMI's requirements, and the functions performed by the two endpoints and the administration server in the agent architecture.

5.1.1 What is an Agent?

An agent in this case is a piece of software that is as much a *software device driver*, as is needed for a PC to talk to its Network Interface Card (NIC), as it is an application. It shares attributes of both.

5.1.1.1 Agent as Software Device Driver

The agent takes its place in the OSI seven-layer model just above layer 2 or the Link Layer and is sometimes referred to as a *shim*. This is because of its very thin nature, reflecting NIMI's demands for it to be *lightweight*. It is also very *portable*, currently developed to cater for 17 different hardware platforms from MVS through to Windows 2000. Each platform has its own agent or *performance endpoint*, but they can all function in the same way.

The agent's role as software device driver in *passive monitoring* is different to that played in active monitoring in that only Windows platforms support the ability to monitor applications in real-time as they leave and re-enter the PC. This is owing to their dependence on interacting with Windows sockets, a software device driver that mediates between the Link Layer 2 and the upper layer protocols.

The agent can report on many more things other than response time. For example, as it sees the Network Layer portion of each packet pass by, it extracts the destination information in a transaction and is therefore able to report on which servers an application has visited *without* putting an agent at the server end. This holds true for complex *data streams*, where a single web query can result in multiple simultaneous requests being sent to one or more servers. It does not hold true, however, for two- and three-tier architectures, where the PC only ever sees, and has contact with, a single server while that server may be talking to one or more other servers in the background.

5.1.1.2 Agent as Application

The agent's role as an application in both *active and passive monitoring* is the same irrespective of platform.

The notion of *synthetic transactions* was introduced in Section 4.3.3 (Strengths of Capture and Playback) as a means of reproducing applications traffic without the need to call on the application itself. This is important for two reasons. First, it is not intruding on the application (as is the case with ARM), and second, operating at Layer 7 of the OSI model, it does not care about any of the underlying protocols or topologies that it runs over.

5.1.2 Endpoint and Server Functions

This section describes the two scenarios of *active* and *passive* monitoring.

5.1.2.1 *Active* Architecture

In *active monitoring*, two endpoints are involved. Endpoint 1 receives test instructions and schedules from the server and is responsible for the delivery of data usually every hour back to the server. Independence from the server is achieved by Endpoint 1's ability to retain data for up to 2 weeks. This allows critical server maintenance to be carried out without compromising performance data and effectively *decentralizes control of the measurements.*

Changes to tests are conveyed to Endpoint 1, which reissues them to Endpoint 2 *dynamically*, without the need to restart the agent.

Lastly, the agents are able to upgrade themselves by contacting the server periodically for new versions, extending their level of independence still further. This has meant that dial-up links can now be monitored as Endpoint 1 can initiate requests for tests. This functionality extends to *passive monitoring* too.

This is perhaps not *self-configuring* but more self-maintaining, which is vital in large-scale deployments.

5.1.2.2 *Passive* Architecture

In *passive monitoring*, only one endpoint agent is necessary at the client end of the transaction to monitor applications. There are advantages to having endpoint agents on the server in that basic system monitoring can then be performed to report back on central processing unit (CPU) utilization (how hard it is working), Input/Output (I/O) volumes (how much data flow in and out of the physical port on the NIC), and memory status or paging (how the memory is coping).

Endpoint-to-server communication is identical to that stated above, in that data can be held at the endpoint for up to 2 weeks, and connectivity can be via a dial-up connection, with the endpoint taking responsibility for requesting upgrades to new agent versions.

Figure 5.1 illustrates the two different scenarios and the role of the administration server.

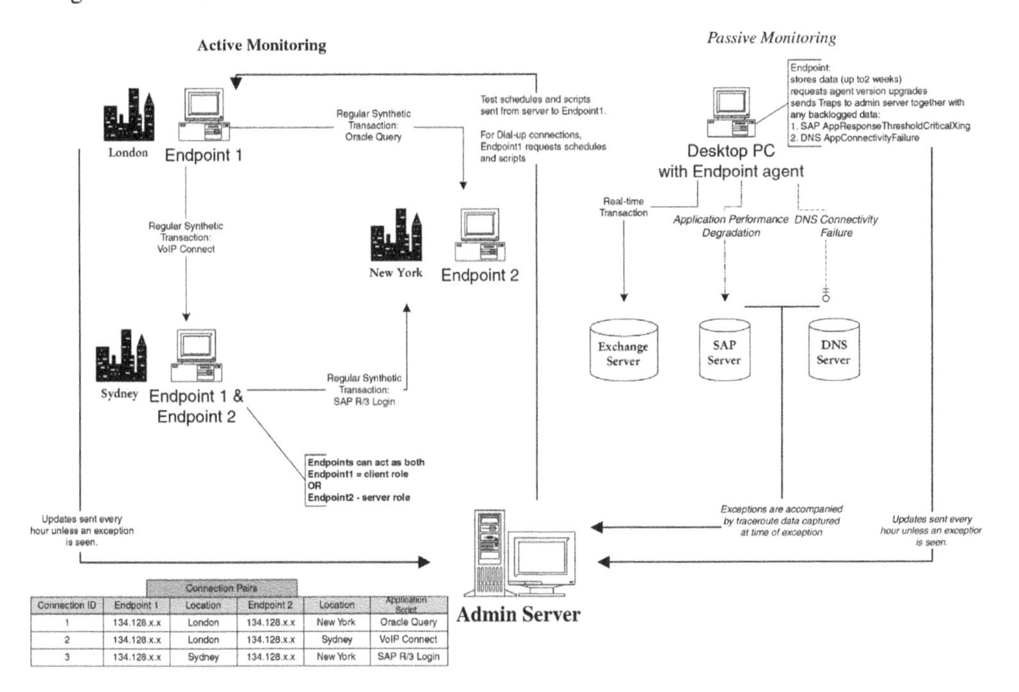

Connection ID	Endpoint 1	Location	Endpoint 2	Location	Application Script
1	134.128.x.x	London	134.128.x.x	New York	Oracle Query
2	134.128.x.x	London	134.128.x.x	Sydney	VoIP Connect
3	134.128.x.x	Sydney	134.128.x.x	New York	SAP R/3 Login

Figure 5.1 Agent architecture showing active and passive scenarios.

5.2 TCP Flows

This section examines more closely the characteristics of TCP flows. These allow the causes of application response time degradation to be identified as resident at the client, in the network or at the server. This ability to deliver more for less more than satisfies the NIMI need for *high yield*.

This breakdown is achieved by looking at the components of a TCP flow, as illustrated in Figure 5.2.

From Figure 5.2, it can be seen that a certain amount of measurable time is spent in both client and server portions of the transaction. It then follows that the remaining time can be attributed to the time that packets spend traversing the network.

The Round Trip Time (RTT) is calculated from the initial setup (Connect) and return ACK when no data are included in the packets. The server and client times are aggregated for a single transaction. In ERP applications, a single transaction can contain tens, hundreds, and even thousands of network turns.

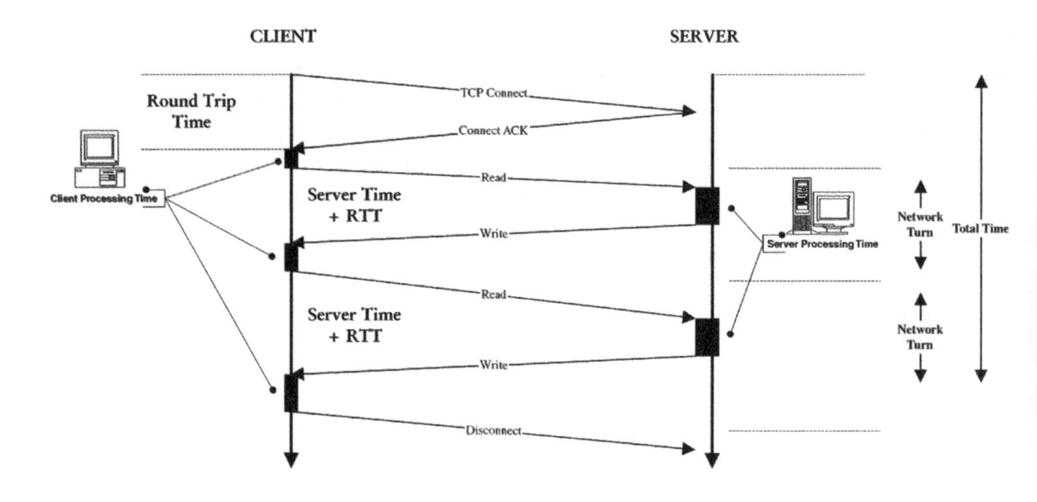

Figure 5.2 TCP connection flows in a simple, single-tier architecture.

While not providing the root cause in every case, it is certainly a pointer to where efforts should be concentrated to find it. In the case of DNS, for example, which underpins every transaction on the network, it is possible to isolate the root cause by capturing when DNS servers are unavailable. If the DNS server does not exist, this would point clearly to a DNS faulty configuration on the client PC.

5.3 Summary

Agent technology of this kind fulfils the aim of supplying real-time application performance monitoring, giving a truly end-to-end view as experienced by the end user. We now go on to look at a viable means of delivering that vital information to aid in root cause analysis.

6

An Integration Model

This chapter reminds us why we need to combine traditional network management with the agent technology seen above to form a powerful partnership and puts forward a means to deliver end-to-end applications monitoring. The study is broken down into three levels in the pursuit of this goal:

- Level 1 – *proactive*, using trap forwarding and correlation to determine possible network failures, application performance and throughput degradation;
- Level 2 – *probable cause diagnoses*, using an example of how this combination can lead to a root cause analysis after only four steps or less;
- Level 3 – *bi-directional interaction*, a plan to link the two systems to enable the drilling down into related *event* data.

The first two levels have been accomplished, and challenges associated with the third are recorded together with possible solutions.

6.1 Why Integrate?

We need to remind ourselves that network management is a means of monitoring and controlling network growth in a sustainable fashion without impacting services or SLAs. To that end, it must be cross-domain, cross-functional, and cross-business group in its makeup and delivery.

A thought, echoed by many in the IT profession, is summed up by Nemzow:

> ...there is no distinction between design of enterprise networks and the applications that overburden them... Application designers forget the impact of routeing, hops and server access when building mission critical software... [19]

6.2 Application-level Traps

This section looks at the role of the applications-level trap and the kinds of traps that are available, linked to the agent technology. A short explanation of automatic thresholds at the application level is given to understand how the agent thresholds move in line with an ever-changing network and how they play an important part in a network management framework.

In the context of a network management framework, SNMP traps were seen to have both a beneficial and potentially detrimental impact on a network in SNMP Traps (Section 2.3.2.1).

It will be remembered from SNMPv2 that manager-to-manager communications made it possible for one NMS gathering data from a set of SNMP agents to forward those data on to other managers.

This particular agent technology employs an SNMPv2 agent and associated MIB to forward traps on the following aspects of applications monitoring. They are sub-divided into *active, passive*, or *both* to make it clear which traps apply to which scenario.

Active monitoring:

- ResponseThresholdCriticalXing – response times between pairs of endpoints is degrading;
- ThroughputThresholdCriticalXing – throughput between pairs of endpoints is degrading;
- ConnectivityFailure – possible network failure between two endpoints.

Both:

- EndpointFailure – server cannot contact an Endpoint1 or has not heard from it after a timeout period.

Passive monitoring:

- AppResponseThresholdCriticalXing – application response time has breached a threshold;
- AppThroughputThresholdCriticalXing – application throughput has fallen below a threshold;
- AppConnectivityFailure – a client application has failed to contact a server[1].

This only represents seven out of 22 traps in total. For every crossing of a threshold, there is an equivalent *NormalXing* to show that normal service has been restored.

6.2.1 Application Thresholds

Thresholds in this instance are not necessarily static values or values that track rising and falling thresholds over time as with MIB-2. If they are set to automatic, they compute an algorithm based on a combination of response time, throughput, and availability. A running 2-week baseline is maintained, against which current values can be compared. This is particularly useful when operating a network where regular changes are made, as it means the thresholds are *self-configuring*.

The only case where one might consider applying a static value would be for the purposes of SLA agreements, but frequent revisits of those values would be necessary if response time were included in the SLA.

Another advantage of the running 2-week threshold is that reports can reflect changes for the better as well as for the worse. We tend to concentrate on the performance, whereas, in fact, it is just as important to register a successful upgrade of a link and to have the data to back it up. Much emphasis is placed on the business justification for such an upgrade, but little is often done after the fact on the business verification of the same case.

In a network management framework, traps and thresholds are crucial.

Traps, once turned into meaningful *events* (see Section 2.3.2.1) that an operator can understand, alert an Operational Systems and Support team (OSS) to a problem ahead of any

[1] This is more likely caused by a server or network problem, since the client sent the trap. This agent makes the distinction between a host that is unreachable and an application that is unavailable on the host.

traditional polling management station. Depending on the size of the network being managed, polling intervals can vary from 3 min on a SME network to 10 min on a national or international carrier network. Those are vital minutes in which to act.

Thresholds, once crossed over a sustained period of time, alert operators to an impending problem that needs to be investigated and confirmed as non-service affecting before being cleared.

When thresholds and *events* are combined from agent technology and passed up to an NMS framework, one can see that the long sought-after proactive approach becomes a reality. We now take just one example of how root cause analysis was made possible through this combination.

6.3 Root Cause Analysis – An Example

This section will trace a real application-related problem from its first appearance on screen through just three steps to finding the root cause.

The scenario is this: an OSS team has a screen that brings all *events* on a network to their attention. The screen is divided into two sections:

- 'broken' browser – made up of critical *events* that need fixing, e.g. hardware failure;
- proactive browser – made up of major and minor *events*.

6.3.1 Case Study

Step 1

A major *event* is generated from an endpoint agent on a desktop PC in Taiwan of the type AppConnectivityFailure for DNS. A web page is brought up to investigate further, as shown in Figure 6.1.

Step 2

Drilling down confirmed it to be an availability issue, as shown in Figure 6.2.

Step 3

A check was performed to see if Taiwan had its own DNS server, and it was found that it depended on one in Singapore. Its hostname was zsngh009.

Step 4

The following pointed clearly to an incorrectly configured DNS entry at the client end (Figure 6.3).

With experienced knowledge of the names of the DNS servers, this process could have been, and in fact was, completed in three steps.

We now look at how I wished to enhance this framework and agent collaboration still further by investigating the possibilities of a return communication from web browser to framework NMS.

6.4 Bi-directional Interaction

The purpose of this section is to outline the plan to take this collaboration one step further. It looks at the transactional breakdown view and investigates ways to provide a dynamic link back to the related source of the degradation in the NMS framework.

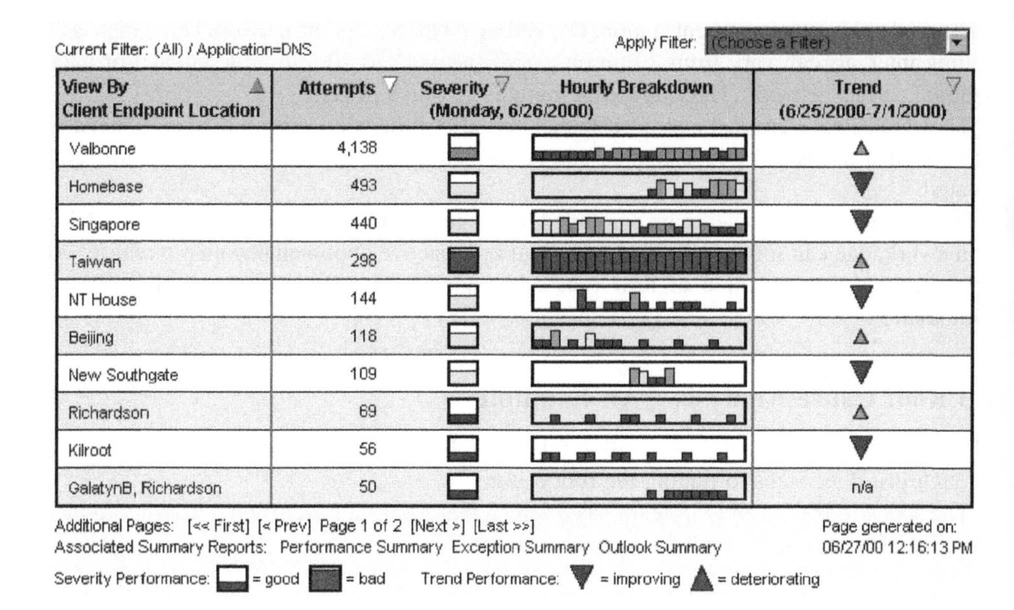

Current Filter: (All) / Application=DNS   Apply Filter: (Choose a Filter)

View By Client Endpoint Location	Attempts	Severity	Hourly Breakdown (Monday, 6/26/2000)	Trend (6/25/2000-7/1/2000)
Valbonne	4,138			▲
Homebase	493			▼
Singapore	440			▼
Taiwan	298			▲
NT House	144			▼
Beijing	118			▲
New Southgate	109			▼
Richardson	69			▲
Kilroot	56			▼
GalatynB, Richardson	50			n/a

Additional Pages: [<< First] [< Prev] Page 1 of 2 [Next >] [Last >>]   Page generated on: 06/27/00 12:16:13 PM
Associated Summary Reports: Performance Summary Exception Summary Outlook Summary
Severity Performance: = good = bad   Trend Performance: ▼ = improving ▲ = deteriorating

Figure 6.1 Step 1. Taiwan appears to have a DNS problem.

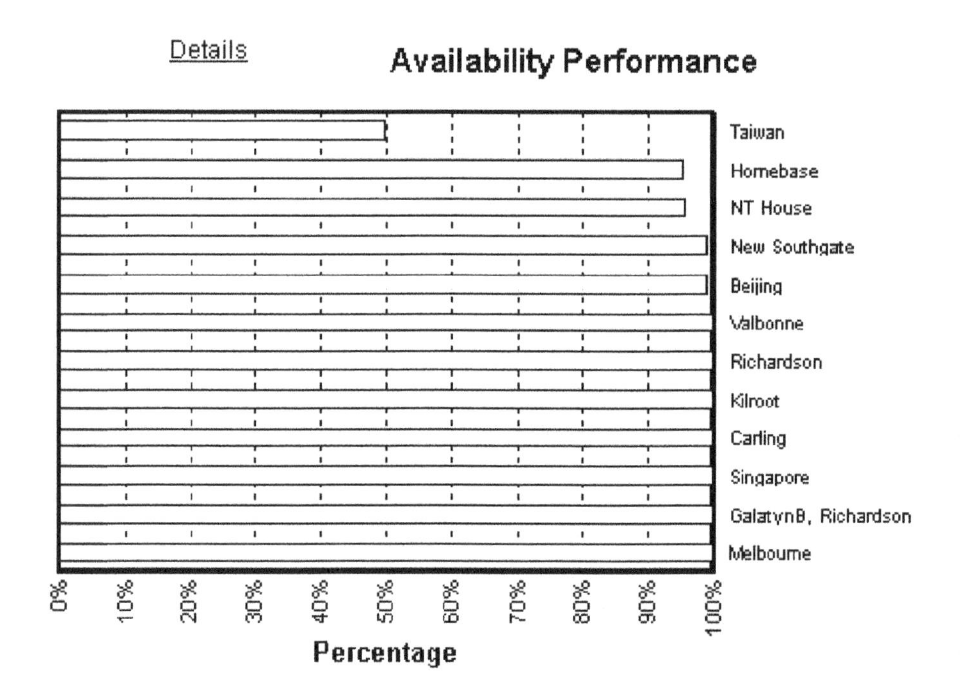

Figure 6.2 Step 2. DNS availability issue confirmed.

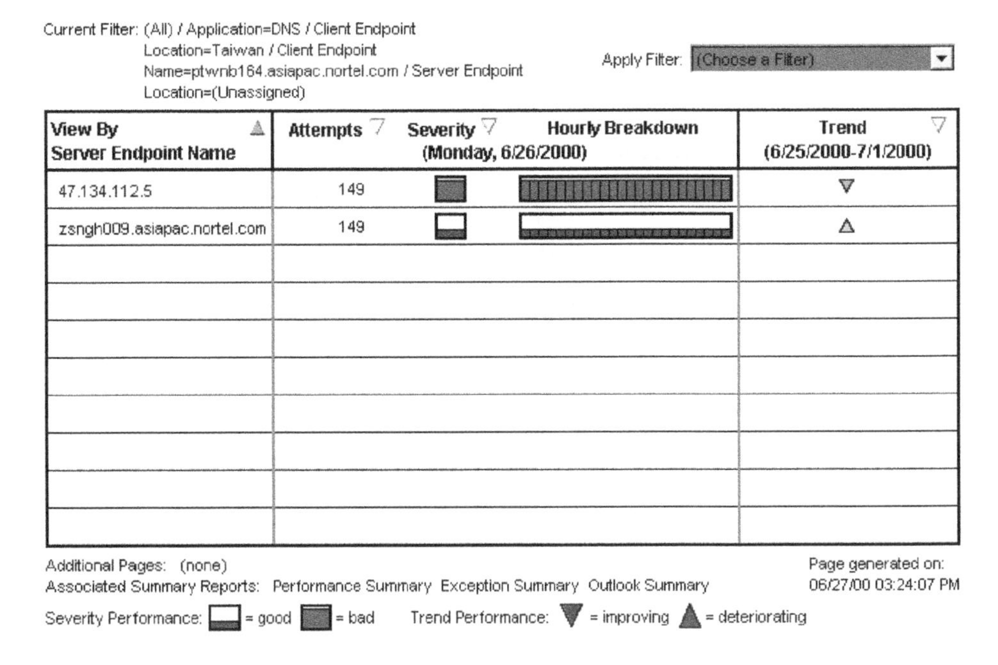

Figure 6.3 Step 3. Root cause: incorrectly configured client.

We also look at the challenges faced and the problems encountered en route to provide a more integrated solution.

6.4.1 Transactional Breakdown

As mentioned above in TCP Flows (Section 5.2), one of the more striking features of the agent's operations is the ability to break down a transaction into time spent at the client, in the network and at the server (Figure 5.2).

These are represented in the web browser, as in Figure 6.4.

Great benefit could be provided by having each portion of the graph above dynamically linked back to the framework system to extract the related *event* data.

6.4.2 Challenges and Possible Solutions

The challenge is to be able to provide linkage from a bar on a graph from one third-party product to interrogate another database on another's.

The possible routes to take in this instance would be perhaps:

- use Perl to post-process the default report and propagate it into another browser window;
- hard-code a link by modifying the report template files;
- use scripts in C to create a new live system effectively, as thresholds are applied before reporting is invoked and are completely independent.

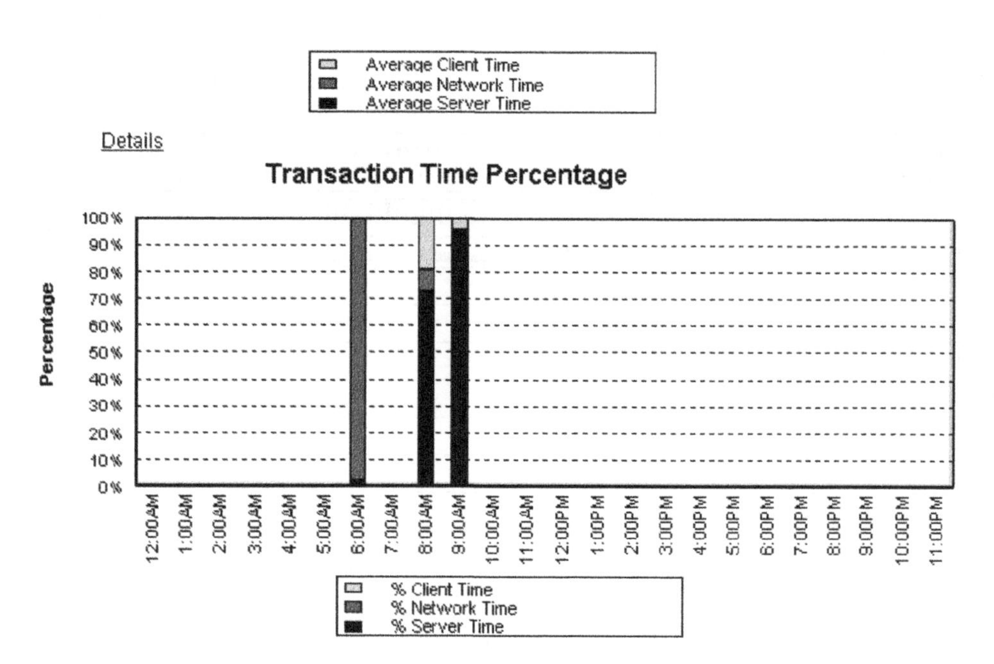

Figure 6.4 Step 4. Representation of client, network and server portions in a transaction.

Because of a lack of programming expertise, the second option was pursued, but this was not without its problems.

Among the first was the realization that the template files were written in a very old version that was not entirely compatible with the latest version of the report package. This meant that I could read some templates but not crucially the template depicting composite transactions times, as shown in Figure 6.4.

Another obstacle was found to be the database of the framework solution, which varied from one of its products to another. In spite of these setbacks, there is scope for further investigation and development for one more familiar and well versed in the gentle art of programming.

7

Application Performance in a Nutshell

Before moving on to examine how service levels can be effected in a move to wireless networks, it is as well to take stock of what we have found to be possible in fixed networks. To that end, this chapter presents a synopsis of the different mechanisms for determining application response time to assist in service-level management and a summary of their contributions in aiding root cause analysis in fixed networks.

7.1 Synopsis of Mechanisms

Historical standards are limited in their capabilities and are of no use in isolation. Each provides the following within the scope of applications' response time measurement and its incorporation into a network management framework:

SNMP v1, 2 and 3 – provide the transport mechanism between managers and agents;
MIB I, MIB II and extensions – provide counters and gauges on single devices;
RMON and RMON2 – monitor whole LAN segments on demand, alarms and events, application-level statistics, 'bandwidth hogs', but no response time from end-user perspective;
CMIS – reports counters and gauges, has a complex OO structure, in which the historical retrieval of data in arrays is nigh impossible.

Current Standards:

SMS – performance monitoring on servers only;
ARM – invasive, not entrusted to critical applications;
CIM – software distribution, inventory, and metering, rather than response-time measurements.

The latest developments can be classified as either invasive or non-invasive. Application instrumentation is considered so invasive as to be detrimental, and network probes, while being non-invasive, are normally expensive to deploy, requiring additional hardware with every change of line speed.

Intelligent agents are thin by nature, require no additional hardware, can be upgraded automatically, and can be either invasive or non-invasive to suit the purpose. In practice, a combination of both delivers regular trend information through active or invasive monitoring and the real-time end-user perspective through passive or non-invasive monitoring.

7.2 Summary

Application response time is important. Ping is no longer sufficient to manage networks whose complexity of construction is now matched by that of its applications traffic.

Surveys indicate that SLAs will live and die by this information in the very near future. What large carriers of voice traffic have been used to and willing to sign up to in the past is driving the present requirement for data to be accountable in the same way. Past standards coupled with innovations in intelligent agents provide a powerful means of delivering on that requirement today.

This does not happen effortlessly. The key is in the integration of such agents and standards into a network management framework: the larger the organization, the higher the level of integration required. A working model of one such integration to assist in root cause analysis has been provided.

The nature of applications changes from quarter to quarter. They have always competed for available bandwidth and by and large have co-existed, albeit in their own chaotic way, which gives a network its 'bursty' tag. In the relatively short period of writing, what was a focus on the response time sensitivity of an ERP application now extends to the data loss, throughput, and synchronized delivery of voice over IP, video-streaming, and interactive TV. The monitoring focus has changed, and at the head of the queue now is QoS, particularly when traversing from fixed to wireless networks, which is where the rest of this book now concentrates its attention.

8

Service Level Quality Across Fixed and Wireless Networks

The second half of this book reviews the architectures planned for second- and third-generation mobile networks, and their ability to provide multimedia services. Crucial to their success is the marrying of two hitherto disparate networks, which are overseen by different standards bodies and different methods for transporting traditional services, and whose customers have been serviced by different suppliers.

To assist in that marriage, some early mandatory standards have galvanized vendors and strategists alike to focus on those elements most likely to deliver voice, data, and video services within a guaranteed and chargeable service level. This document researches those elements and provides empirical data of how sample applications perform in a simulated wireless environment and how they react when competing with other applications.

8.1 Relevance

The driving force behind the activities referred to above is the business, with technology as the enabler. The attraction of offering a Premium service in any industry sector is very compelling, but to offer it to 30 million subscribers of mobile phones, set to rise to 1.3 billion by 2004 (In-Stat Research), is too good an opportunity to miss.

What have been missing, however, are the empirical data to indicate that those premium services are feasible and that the level of delay will not be unacceptable to potential customers, effectively killing the goose before it has had a chance to lay any eggs of any description.

8.2 Scope

The following areas are covered:

- an overview of wireless architectures and mandated standards;
- a review of components both on fixed and wireless sides, which could assist or hinder application performance when moving to wireless;
- an assessment of previous research passing Web traffic over mobile networks;
- application performance in a simulated wireless environment.

8.3 Summary

In short, applications will need to be frugal with their resource consumption with a balance struck between compression techniques and the probability of lost or delayed data.

There is increased scope for delay variation and interruption to service, which will make services in many cases as unreliable as those experienced on mobile networks today.The handling of radio resources and contention management techniques will be key to the delivery of an effective QoS.

9

Wireless Architectures

This chapter gives brief overviews of second- and third-generation mobile networks as an introduction to how data traffic is being introduced progressively into voice networks either as an overlay network (i.e. one running on top of or in parallel with another) or as an integrated multiservice[1] network. In the case of third-generation networks, a summary of the Third Generation Partnership Project (3GPP) mandated standards and their implications for delivering a guaranteed service level is included.

The following areas are examined:

- Global System for Mobile Communications (GSM) or 2G networks;
- General Packet Radio Service (GPRS) or 2.5G networks;
- Universal Mobile Telecommunications System (UMTS) or 3G networks;
- Implications of one 3GPP mandated standard.

9.1 Summary

Second-generation (2G) wireless networks use circuit-switched technology based on Time Division Multiple Access (TDMA) methods and are designed to support low-speed data services, to be compatible with Integrated Services Digital Network (ISDN), and, most importantly, to support international roaming. Pre-2000 figures state that 140 operators in 80 countries worldwide had adopted GSM, representing over 85% of all subscribers for digital communications. A threefold increase in user capacity moving from Frequency Division Multiple Access (FDMA) analogue technology to TDMA[2] was matched by another threefold increase with the introduction of the Digital Cellular System (DCS1800) operating at a higher radio frequency range (1800 MHz). This is often referred to as the Personal Communications Network (PCN).

The high costs for circuit-switched GSM connections and a paltry 9.6 kbps data rate deterred all but the most hardy from attempting to use the data service. With the advent of web-based applications, the need for a more cost-effective, efficient, and reliable means of transferring data was required. This optimized access to Internet applications, which we know

[1] A multiservice network is taken to mean the delivery of voice, data, video, and audio traffic over the same network infrastructure.

[2] A description of TDMA, its evolution from FDMA and possible migration to CDMA follows in Section 2.2.2.1.

as GPRS [20], was overlaid on to GSM networks using packet-switched technology and provides theoretical data rates up to 160 kbps.

As audio and video compression techniques improved, the need for an entirely packet-switched network (effectively an extension to the Internet) with increased capacity to support multimedia services was realized in the third-generation (3G) specifications. Delays in delivering networks and services have turned market attention to the here-and-now technology that is GPRS.

9.2 GSM (or 2G) Networks

This section looks at the structure of GSM networks, their basic components, and their characteristics to gain an appreciation of how wireless differs from fixed (sometimes known as wireline) networks and of how the inherent difficulties affect performance guarantees over such networks.

In particular, we will look at each of the following in turn:

- the architecture of GSM and its Network and Base Station subsystems;
- the Radio Link function;
- the Network function;
- the integrated voice and data services supplied.

It will be seen how the inevitable instabilities of a terminal 'on the move' contribute to a quoted [21] packet loss of 5–15%, latency of 100–300 ms and jitter[3] rates of 50% on an audio stream transmitted at just 13 kbps. For some sensitive multimedia applications such as Voice-over-IP (VoIP) it has been shown that near-toll quality voice can only be achieved with 0% packet loss and jitter rates of less than 2% [22], which equates to less than 5-ms variation in the arrival rate of voice-packet data.

9.2.1 GSM Architecture

The GSM architecture can be divided into three areas:

- Mobile Station (MS) – a handset or terminal, traditionally a mobile phone, but now can mean any GSM-enabled device, e.g. Personal Digital Assistant (PDA) with a GSM 'cradle'[4];
- Base Station Subsystem (BSS) – controls the radio link to the MS and consists of a Base Transceiver Station (BTS) and a Base Station Controller (BSC);
- Network Subsystem (NSS) – controls mobility management (MM) as a user moves as well as controlling connections to other fixed or mobile networks.

The mobile station holds a smart card within it called the Subscriber Identity Module (SIM). This holds the International Mobile Subscriber Identity (IMSI) used to identify the subscriber and a secret key for authentication. A copy of this key is stored in the Authentication Centre (AuC), one of four databases controlled by the Mobile Services Switching Centre

---

[3] 'Jitter' is the variation in latency (or packet arrival rate) that results in poor voice and video quality.

[4] A 'cradle' is a hardware extension snapped into place on the back of a PDA. Laptops also can be equipped with GSM PCMCIA cards with similar functionality.

(MSC), and is used for encryption over the radio channel as well as authentication. The IMSI identity is stored in the Home Location Register (HLR) database of the subscriber's service provider and transferred into the Visitor Location Register when the mobile station is roaming between countries or between service providers.

The International Mobile Equipment Identity (IMEI) identifies the unique handset number, which is stored in the Equipment Identity Register (EIR). This ensures that stolen or non-approved handsets do not gain access to the network.

This architecture is represented in Figure 9.1.

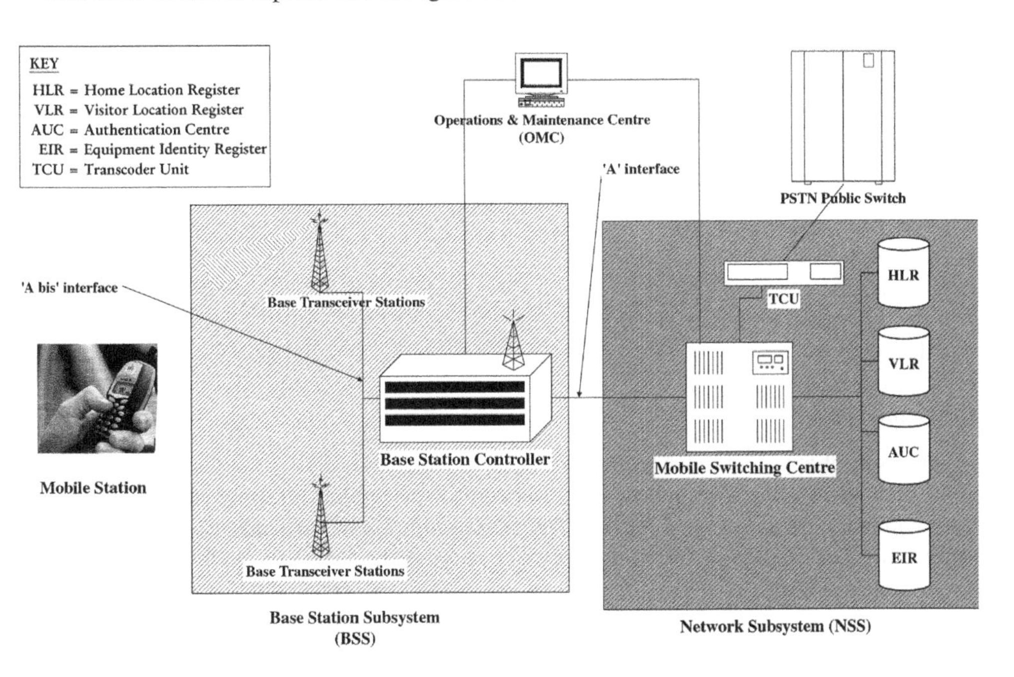

Figure 9.1 GSM architecture.

In the BSS, the BTS handles all radio link protocols and cell definition, communicating to the BSC via a standardized link: the A bis interface. All links or interfaces between elements in a wireless network are given a two-letter identifier. The BSC manages the radio resources (RR) for one or more BTSs, including channel setup, frequency hopping[5], and handoffs[6] between the BTSs under its control.

The Network Subsystem comprises an Operations and Maintenance Centre (OMC), from which network management, administration, and control of the ground components of the BSS

---

[5] Frequency hopping is a technique used in TDMA to transmit each timeslot in a different frequency. This reduces co-channel interference, alleviates multipath fading and is viewed as another security measure against eavesdroppers. We will return to study these obstacles to performance in Section 9.2.2.1 Noise and Interference.

[6] Passing control from one BTS-controlled cell to another or changing frequencies within the same cell as a result of low signal strength, interference, or bit error rates (BER) being too high. Signal strength can be monitored from the BSC or, in TDMA networks, can be combined with reports from the mobile station, known as Mobile Assisted Handoff (MAHO).

can be performed, and the main Mobile services Switching Centre (MSC). The latter is like any normal Public Switched Telephone Network (PSTN) or ISDN switch with the added functionality of being able to handle mobile subscribers. It houses the databases governing registration, authentication, location updates, and handoffs as well as call routeing to roaming subscribers.

The Transcoder Unit (TCU) converts normal 64-kbps speech, managing to squeeze (or multiplex) four voice channels (or conversations) into a single 16-kbps channel, which is the standard in GSM systems.

9.2.2 The Radio Link Function

This is by far the most important aspect when contemplating guarantees on service level performance. For those unfamiliar with wireless transmission and its difficulties, it is worth digressing to consider what those issues are before seeing how these have been overcome or diminished by the countermeasure techniques that exist today. The issues include:

- noise and interference
- free space loss
- attenuation
- reflection and multipath
- delayed spread
- Rayleigh fading
- Doppler shifts

The result of applying those techniques is a service guaranteeing a total usable bandwidth of 13 kbps for voice traffic, power emission controls to reduce environmental concerns, an imaginative use of silence, and reduced power consumption in the handsets, which lengthens battery life. The challenge now is not only to guarantee the same service levels for data, but also to extend them.

9.2.2.1 Noise and Interference

We must first understand how channels are created. The International Telecommunications Union (ITU) responsible for radio-spectrum allocation designated the frequency bands for GSM mobile networks in Europe:

- 890–915 MHz for uplink communications from mobile station to BTS;
- 935–960 MHz for downlink communications from BTS to mobile station.

Later, spectra in the 1800-MHz and 1900-MHz range were allocated because of capacity demand. These gave rise to the service names DCS1800 and Personal Communications Service PCS1900 (in the US only).

The 25-MHz bands above were divided into 125 channels (124 channels and one control channel), with each channel having a bandwidth of 200 kHz. These channels are then subdivided into eight timeslots (TS) each of which can be used to carry voice traffic. This use of timeslots is the basis of TDMA technology and is described in Figure 9.2.

We will return to the TDMA frame structure again (Section 9.3.2 GPRS Lower Layer Protocols) when considering how GPRS evolved from this model to enable it to carry larger amounts of data.

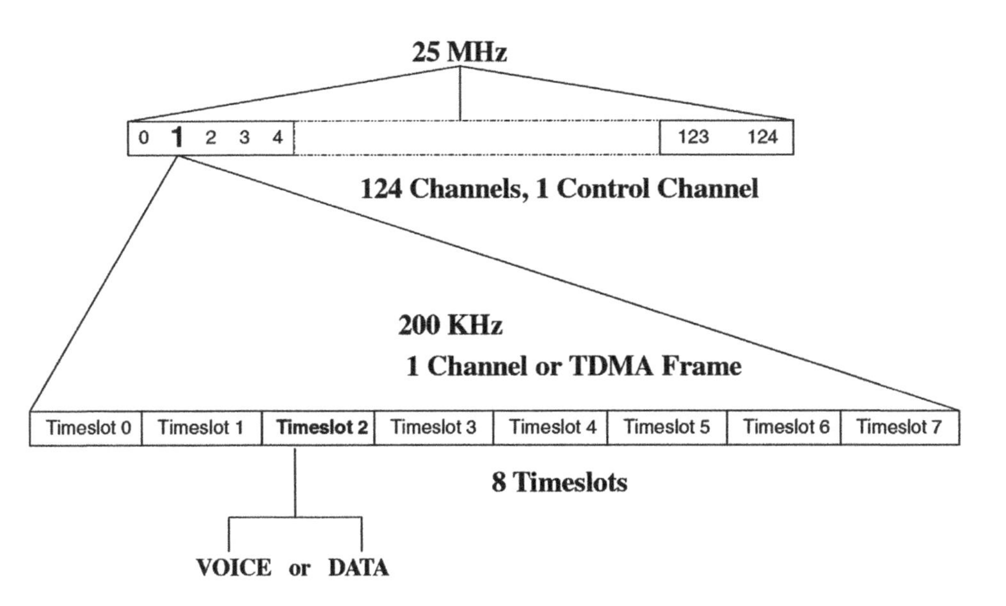

Figure 9.2 TDMA radio access method.

The great advance that this has over the previous FDMA technique is that no one channel is dedicated to a subscriber. Moreover, voice traffic is transmitted in short bursts in assigned timeslots, reducing the likelihood of *co-channel interference* and enabling frequencies to be re-used more often. This leads inevitably to the capacity to service more subscribers.

Remember that a radio signal is a wave of electromagnetic energy propagated through space and that this is subject to distortion by other sources of the same energy. Lighting and ignition from spark plugs of a car are two familiar examples. Anyone placing a mobile phone near a digital phone will have noticed similar disturbances to the airwaves, creating *noise* that sounds like static.

Interference is of two types: *adjacent channel interference* caused by one channel spilling over into another and *co-channel interference*, which occurs when two signals are transmitted on the same frequency and are received by the same receiver (BTS), superimposing one signal on the other and rendering both unrecognizable.

The cure for the first type is to have a 'guard band' between the two channels, which effectively segregates them using unused bandwidth. The second can be avoided only through careful planning of cells and their allocated frequencies.

Figure 9.3 shows, in frequency waveform terms, how the different interference patterns discussed in this section can be distinguished and details some of the countermeasures that can be taken to combat them.

9.2.2.2 Free Space Loss

This is simply the condition that the further from the source transmitter you are, the weaker the signal will be. This is known as the *inverse square propagation law*, in which, under perfect conditions in a vacuum state, the power output diminishes by the reciprocal of the square of the distance travelled. Some examples are given in Figure 9.4.

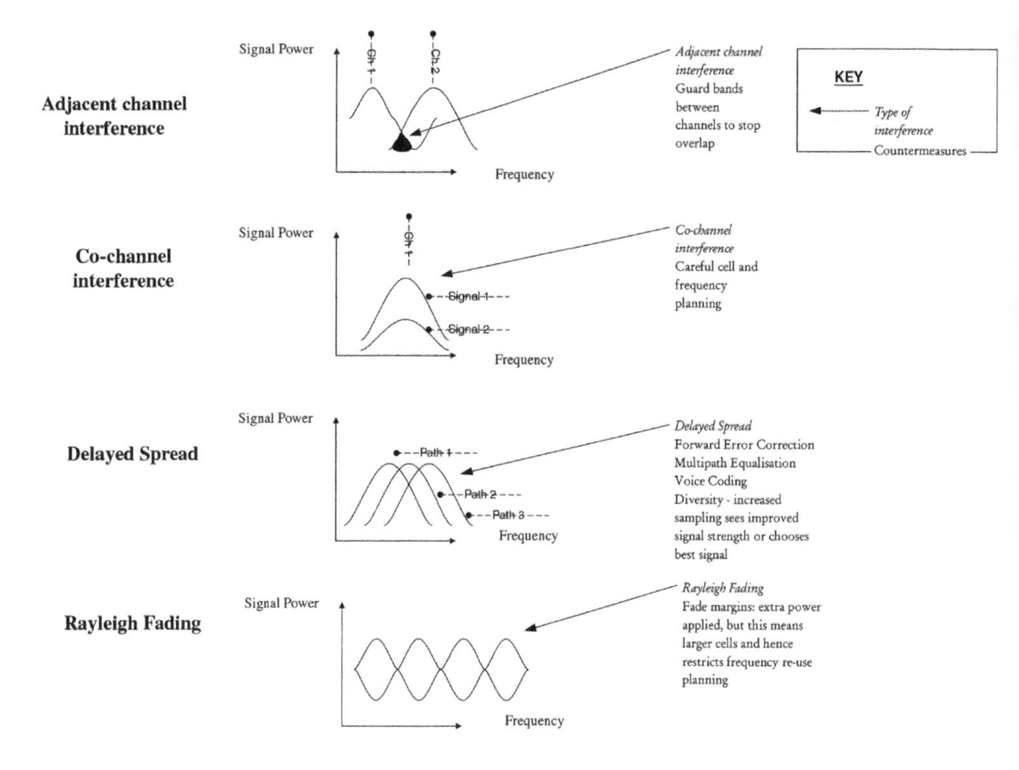

Figure 9.3 Waveform interference patterns and their countermeasures.

Naturally, in the real world, terrain and atmospheric conditions play their part, reducing the power more markedly as does the frequency of the transmission. The lower the frequency (below 1 GHz), the more resilient it is to those conditions, which is why terrestrial microwave transmitters (above 10 GHz) and laser systems require line of sight and can be affected by fog and snow. Broadcasting frequencies, at around 4 GHz, are less affected by fog but can degrade in heavy storms.

A more pictorial, real-world representation of the various kinds of interference to be discussed from here is given in Figure 9.4, which gives details of the propagation issues, their interference type, and the advantages and disadvantages they present.

9.2.2.3 Attenuation

This happens when the signal is absorbed or partially blocked by physical features. As discussed earlier, frequencies above 10 GHz are severely affected by poor weather conditions, and those above 30 GHz are not suitable for transmission over long distances.

In built-up areas, attenuation is experienced more, and tall buildings often create *radio shadows*. Entering or exiting a shadow can take some time in a moving vehicle, which is why this phenomenon is sometimes referred to as *slow fading*.

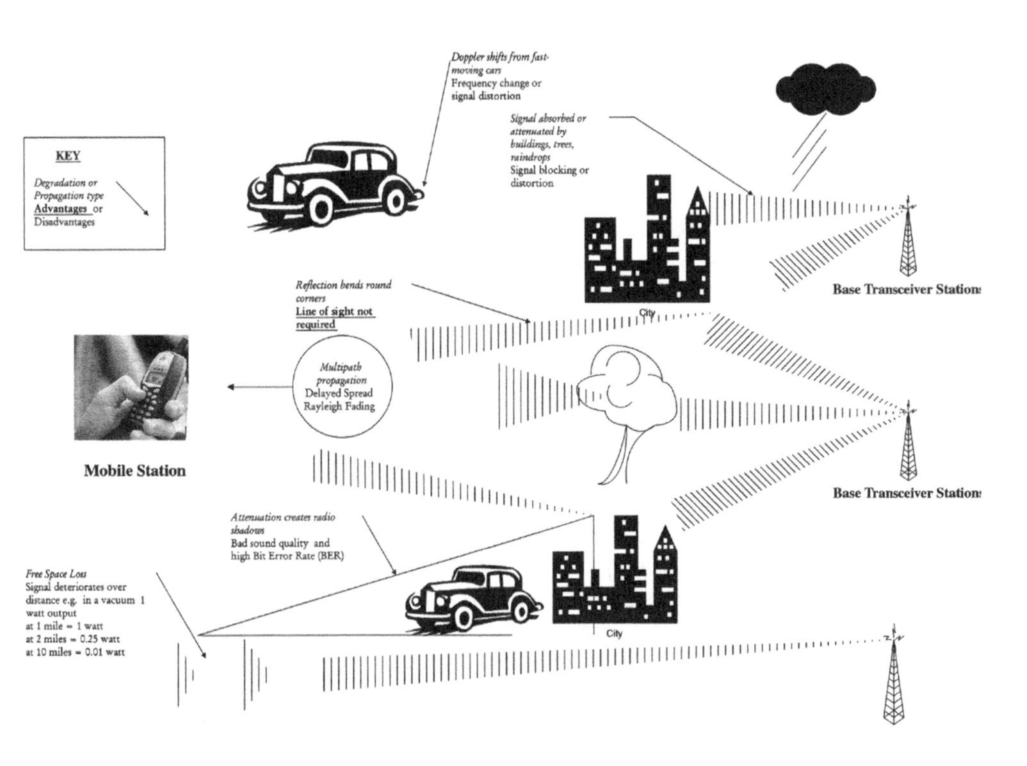

Figure 9.4 Wireless propagation issues.

9.2.2.4 Reflection, Multipath, Delayed Spread, and Rayleigh Fading

Reflection of the signal off buildings can have a positive effect, however, by appearing to bend the signal around corners. This can mean that fewer transmitters have to be deployed. The downside to this is that signals bouncing off several features can result in multiple paths or *multipaths* being taken, each of which can arrive at its destination at different times. The delay encountered is known as *delayed spread*. When two signals arrive in apposition, that is when the peak of one wave is mirrored by the trough of another, this cancels out the signal completely and is known as *Rayleigh fading*.

9.2.2.5 Doppler Shifts

Like the changed appearance of car wheels (or the older variety in horse-drawn carriages) spinning at a certain speed, transmitting from a vehicle at a certain speed may change the frequency of the signal and distort it.

9.2.2.6 Countermeasure Techniques

Countermeasures for most of the degradation issues cited above have been developed with some measure of success. Principle amongst these is *speech coding*. Speech must be digitized, that is sampled at 8000 samples per second and turned into a series of 1s and 0s, before being superimposed on a carrier wave (i.e. the frequency) using a technique called *modulation*. Digital modulation carries out discrete changes to that wave, which can be interpreted by the receiver. It is beyond the scope of this study to examine any of the following modulation techniques in detail. They are listed here for completeness:

- Amplitude Shift Key (ASK) – pulses of ON and OFF relating to 1s and 0 s;
- Frequency Shift Key (FSK) – two different frequencies used for 1s and 0 s;
- Phase Shift Keying (PSK) – uses phase differences of two out-of-phase signals;
- Differential Quadrature PSK (DQPSK);
- Minimum Shift Keying (MSK);
- Gaussian Minimum Shift Keying (GMSK).

The method used for speech coding in GSM networks is *Regular Pulse Excited– Linear Predictive Coder*with a *Long Term Predictor Loop* (RPE-LPC). This uses previous voice samples to predict the current sample and produces better voice clarity and a total bit rate of 13 kbps.

Noise is reduced by the combination of channel coding and modulation to ensure low bit error rates (BERs). This is accomplished by block interleaving and through the coding scheme used for GSM, which is known as *convolutional encoding*. We will revisit blocks and coding schemes when looking at GPRS (Section 9.3.5 Coding Schemes)

As discussed earlier (Section 9.2.2.1 Noise and Interference), co-channel interference is dealt with largely by TDMA's ability to send voice in short bursts and at different frequencies.

Free space loss is offset by the careful planning of cells, which hitherto has been an art all of its own. The next generation of Code Division Multiple Access (CDMA) technology does away with cell-planning headaches, as all subscribers in a cell have their frequencies coded into a 1.25-MHz channel, providing a 10-fold increase in capacity. Remote areas, however, may still encounter free space loss where full coverage is not guaranteed.

Reflection, multipath, and delayed spread are tackled by an individual supplier's *Multipath Equalization*technique, which extracts the best signal and excludes reflections. In some cases, the signals can be combined to produce a stronger signal. This is not covered by the GSM standard. Some alleviation of multipath fading and co-channel interference is also carried out through *frequency hopping*, whereby each timeslot is transmitted at a different frequency.

Methods such as *Discontinuous Reception* (DRX) and *Discontinuous Transmission* (DTX) put the mobile station into sleep mode, thus extending battery life. In the case of DTX, it is estimated that 40% of our time on the phone is spent not talking. By turning the transmitter off during these silences, co-channel interference is minimized, and the subscriber capacity of any given cell is increased. Of course, this requires a Voice Activity detector that is sufficiently sensitive to distinguish between a human voice and extraneous noise. Such devices have been able to distinguish between voice, voicemail, answer machine, and fax for some while now.

Power control generally is handled by the BSC, but it does take measurements from the MS on signal strength and/or quality in terms of BER before stepping the power up or down.

We have seen the measures that can be applied to help avoid disruptions to service associated with the Radio Link function, but these should not be thought of as hard and fast guarantees of service. That requires another set of techniques that will be addressed in Section 10 Moving to Wireless.

Let us now take a look at that part of the network responsible for ensuring that wireless service is continuous once on the move and can be accessed from anywhere internationally: the Network Function.

9.2.3 The Network Function

The issues with wireless transmission referred to above, with the exception of Doppler shifts, exist to or from any stationary MS, whether it is transmitting or receiving. Once that MS begins to move, continuous connectivity and access to services from anywhere are handled by the network function.

It is important to understand the concepts of:

- Radio Resources (RR) Management
- Mobility Management (MM)
- Connection Management

as they control the setup and maintenance of radio channels. It is the allocation and maintenance of these channels that will enable more data to be carried in GPRS networks. We will examine these in turn to see how handoffs are managed, how the network function always knows where the MS is, who is operating it, and what services it has access to.

9.2.3.1 Radio Resource Management

This layer of management controls the setup, maintenance, and termination of radio channels, including handoffs between the MSC the BSS and the MS. It also governs power control, discontinuous transmission and reception (DRX and DTX), and timing advance.

Radio resources are not dedicated for the duration of a call but are released and re-assigned to other users entering the geographical area governed by a single BTS, known as a *cell.* Nor do you have to leave a BTS cell to have your radio resource changed. A high BER can cause a channel re-assignment within the same cell. The switching of radio resources of a call to another cell or channel is called a *handoff.*

There are two types of handoffs:

- internal handoffs – transfer of channels within the same BTS cell or transfer between BTS cells under the control of the *same* BSC;
- external handoffs – transfer between BTS cells under the control of the *different* BSCs or under the control of different MSCs.

The figure below (Figure 9.5) describes the two scenarios. It is worth noting that in the case of the external handoff between MSCs, the original 'anchor' MSC remains responsible for most call-related functions.

9.2.3.2 Mobility Management

This aspect tracks the *location* of the subscriber once the MS has been powered on so that calls can be routed to the correct cell, as well as providing *authentication* and *security* measures.

Details of the mobile user's location are held in two database registers, the HLR and the VLR, both resident in the MSC (see Figure 9.1 GSM Architecture) to enable location update and call routeing to take place.

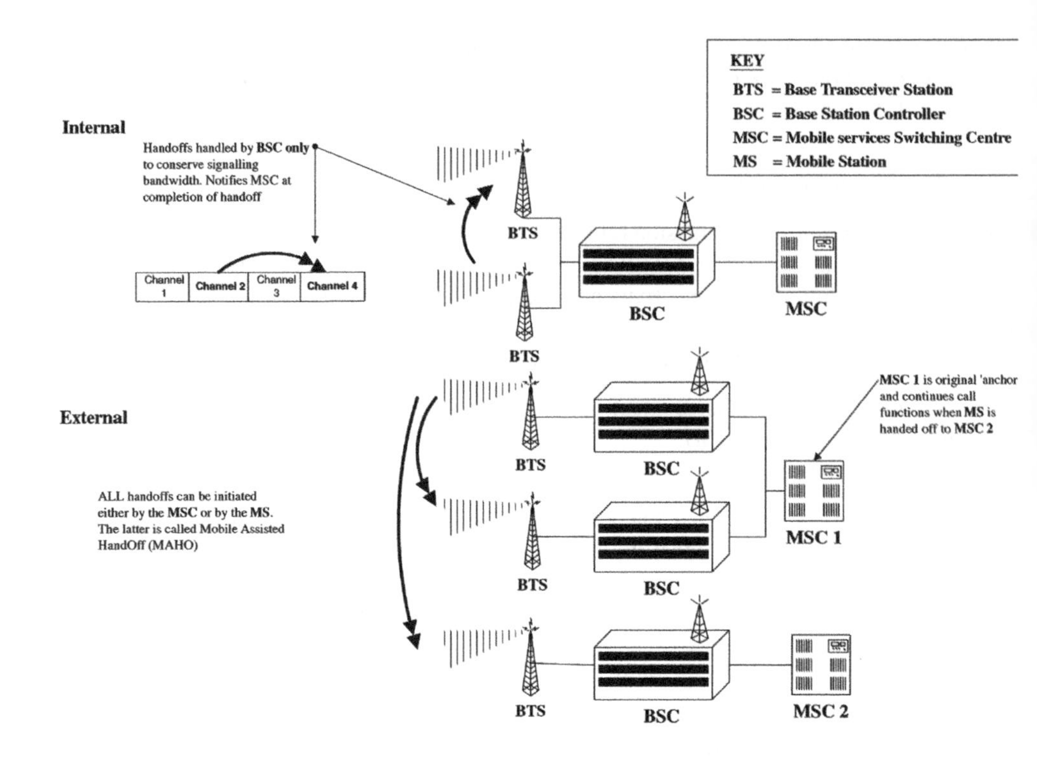

Figure 9.5 Internal and external handoffs.

Incoming calls are 'paged' to the MS by means of a special control channel called a paging channel. The way in which this is done is not particularly efficient in GSM as it uses one of two methods of achieving this. GSM either pages all the mobile stations in the cell or instructs all the mobile stations within a cell to send location update messages to the MSC. There are other special channels that are not data or voice traffic channels, and these are listed below with their functions. This is the reason why the apparent 16-kbps standard bandwidth specification in GSM is limited to only 13 kbps. Some timeslots have to carry control functions for GSM to operate effectively. The types of channel are:

- Traffic Channels (TCH) – that carry voice or data traffic
- Control Channels – used by the BSC to communicate with the MS, including controlling certain functions of the MS. These fall into the following categories:
- Broadcast Control Channel (BCCH) – continual broadcast of base station identity, frequency allocations and frequency hopping sequences;
- Frequency Correction Channel (FCCH) and Synchronization Channel (SCH) – each cell has a timeslot structure to which the MS must be synchronized to preserve the characteristics of a voice pattern. Each cell broadcasts one SCH and one FCCH.
- Random Access Channel (RACH) – used by the MS to make calls or request a service;
- Paging Channel (PCH) – incoming call or service notification from BSC to MS;
- Access Grant Channel (AGCH) – response by the BSC to an RACH request, allocating a channel to the MS for signalling.

There are several levels of *authentication* and *security*. First, and as previously stated, the mobile equipment has a unique identifier (IMEI number in Section 9.2.1 GSM Architecture), which is checked against the EIR database. The response can be one of three:

- *White-listed* – MS is allowed to gain access to the network;
- *Grey-listed*– MS is under observation for possible problems;
- *Black-listed* – MS is non-approved for GSM networks or is stolen and is refused access to the network.

Second, the SIM card in the mobile station stores a secret key that is matched against the AuC database held within the MSC. Third, the use of TDMA framing to digitize and transmit the signal gives protection from eavesdroppers. Additional options exist to encipher the transmission to give enhanced security.

9.2.3.3 Connection Management

Sometimes referred to as *communications management*(CM), this layer is responsible for call control, supplementary services management, and Short Message Service (SMS) management.

Part of that call control relates to call routeing. The directory number issued to mobile phones is called the Mobile Subscriber ISDN (MSISDN) and is defined by the E.164 numbering plan used internationally. That number includes:

- Country code – identifying the home country;
- National Destination code – identifying the subscriber's operator.

The next few digits may point to the subscriber's HLR within the operator's network, which is more formally known as a Public Land Mobile Network (PLMN). This permits the GSM network to locate the subscriber and obtain routeing information from the HLR. It is worth noting at this point that an inefficiency of the GSM network is the long distance route that a call must take when roaming outside a user's home country. The 'triangle' issue arises because the call must be routed to its HLR located in the home country first before being passed on to its destination.

9.2.4 Integrated Voice and Data Services

It is important not to overlook the fact that GSM gave us our first taste of passing data over mobile networks, albeit at a painfully slow rate of 9.6 kbps. Modem-less transmissions became a reality with the introduction of the Personal Computer Memory Card International Association (PCMCIA) cards designed for use in laptop computers hooked up to a GSM phone, and Fax messages conforming to the Group 3 standard could now be sent anywhere in the world from a GSM phone.

Other data services include the vastly underestimated SMS service, which allows the user to send up to 160 alphanumeric characters between mobile phones and store them off-line if the receiving MS is not switched on. The coded language many of us used in the absence of any formal shorthand training has now become commonplace, especially capturing the imaginations of young people. It should have been no surprise to operators as to the impact that secret codes, brevity of communication, and anonymity would have on pupils and students alike.

Not to be excluded is the *Cell Broadcast*, which allows brief messages to be sent to all phones in a specific area. This has applications such as storm warnings, traffic reports, and accident alerts. Although not heavily used currently, we will see more of these location-based services that will come with 3G systems to bring information to the subscriber relevant to their needs.

It is our appetite for better services and faster speeds that prompted operators to look at enhancements to the existing standard to meet medium-term, projected requirements and propose new standards where those services' requirements could not be fulfilled by current technology. The need to evolve was met first with the introduction of GPRS.

9.3 GPRS (or 2.5G) Networks

This section summarizes the enhancements made to the GSM network to enable it to support a range of IP applications and reviews the first attempts to distinguish between different classes of traffic, which would allow some level of performance to be maintained, if not guaranteed.

To study this more closely, we will need to consider the following:

- GPRS architecture
- the associated lower layer protocols
- call setup
- the air interface
- coding schemes
- multislot capability
- QoS profiles

It will be seen how the GPRS network simply introduces an overlay of packet switching, retaining the same cell layout, frequency planning, and TDMA frame structures and channels as the current GSM network. The packet switching element is achieved through the introduction of two new nodes (or pieces of hardware), which provides connectivity from the Base Station Subsystem (BSS) to other packet-based IP or X.25 networks, and these come with their own set of protocols specific to the wireless environment. In order to be able to connect to both circuit-switched voice and packet-switched data separately or simultaneously as the GPRS standard [20] dictates, this requires a change of handset. Forcing users to buy another phone will almost certainly delay the initial take up of IP services until the price matches more closely those of GSM phones.

We will also see that GPRS is effectively a halfway house in its support of IP applications and why a third generation was deemed necessary to support the types of multimedia applications available across the Internet today and those of the future.

9.3.1 GPRS Architecture

We have already seen how the GSM network tripled its capacity and increased flexibility through its use of TDMA technology, but essentially, it is dependent on circuits being set up that assign timeslots for the duration of the call. Packet switching introduces a finer granularity whereby radio resources are assigned to a MS *temporarily on a per-packet basis*. As a result, the likelihood of interference from other frequencies or channels is greatly reduced, and GPRS is much more able to expand to meet the needs of so-called 'bursty'[7] applications.

GPRS was designed to communicate with IP and X.25[8] data networks, these two being the most prevalent in the early 1990s when discussions first began.

At that time, it was the financial service institutions, a large user of X.25 networks for transaction processing, credit checks, and transfers, that were driving demand. One of the largest X.25 global networks is still used for airline reservations and bookings. These networks may have evolved to larger-capacity versions, such as Frame Relay up to 2 Mbps and Asynchronous Transfer Mode (ATM) from 34 Mbps up to 10 Gbps, but they have their roots in X.25 as a verifiable means of ensuring that all the data that leave A arrive safely at B. One might view this as a simple concept, but one that the Internet challenged with its use of unreliable transport protocols, such as User Datagram Protocol (UDP).

The notion of reliability and timeliness lies at the very heart of guaranteeing performance and, as we will see (Section 10.10 Compression standards for multimedia applications), the protocol used for delivering multimedia applications, the Real Time Protocol (RTP) had another protocol designed (Real Time Streaming Protocol RTSP) to accommodate the time-stamps and sequence numbering lacking in UDP.

In order to connect to those networks, GPRS introduces two new nodes: the Serving GPRS Support Node (SGSN) and the Gateway GPRS Support Node (GGSN) as shown in Figure 9.6 with connecting interfaces [23]. For our purposes, we will concentrate only on the shaded area.

The *GGSN* is seen from a neighbouring IP network as an IP router connecting to all MS IP addresses and, as such, may contain firewall, packet filtering, or QoS mechanisms. The GGSN also assigns the correct SGSN to a MS based on location signalled to it by the HLR.

The *SGSN* links the GPRS core network to the BSS, which can be referred to as the Radio Access Network (RAN). With a link into the HLR, the SGSN is able to track the movements of the MS, update its whereabouts and direct data packets to the correct BSS for that location. It also carries out authentication, ciphering, session and logical link management in addition to the mobility management described earlier.

The other palpable change is to the MS or handset itself. In order to receive IP packet-based services such as e-mail and web-based applications, the mobile device must understand the GPRS way of working. The rules that govern any system's *modus operandi* are known as *protocols*, and because different protocols provide different functions, these are *stacked* one upon another. Any MS wishing to join a GPRS network must therefore come with its own *GPRS protocol stack.* GSM mobile phones do not have this capability and cannot be upgraded to support GPRS. The expense of having to purchase a new phone may limit the uptake of such services in the short term.

It should be noted that an additional Packet Control Unit (PCU) has been added to set up, maintain, and disconnect (or tear down) packet-based calls. This can sit physically at either the BSC or BTS, or it can be co-located with the SGSN. The BSC has been enhanced to

[7] We will see how it is not the application that is bursty, but the nature of TCP operations, which expands to fill any given bandwidth resource if left unfettered (see Section 10.3 TCP Operations).

[8] X.25 is an ITU standard defining how connections operate at the edge of a connection-orientated data network. Another related protocol, X.75, defines operations across the core of the same network.

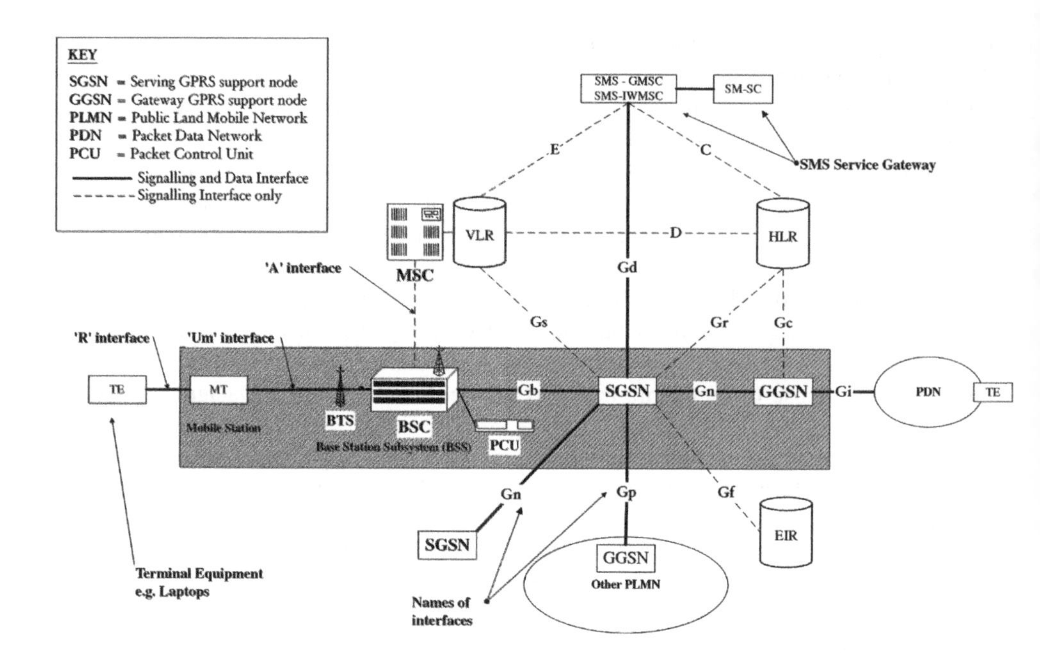

Figure 9.6 GPRS architecture.

support all GPRS lower layer protocols to transfer data over the air interface to the MS. The BTS has no knowledge of these protocols, whose functions we now consider in the following section.

9.3.2 GPRS Lower Layer Protocols

As we will see there are some similarities with the ISO seven-layer model in its use of naming conventions (see protocols LLC and MAC below), but the functions are adapted to suit the radio link connectivity. It is important to note the error correction techniques below, as these ultimately will affect performance either beneficially by ensuring that data arrive in the correct sequence and with the correct content or adversely by introducing delays through excessive amounts of checking.

At the network layer level, IP is used between the SGSN and GGSN. The Gn and Gp interfaces that connect SGSNs to GGSNs are sometimes known as the *GPRS backbone*. So as to accommodate X.25 data packets and IP packets as well as to make the design future-proof, a tunnelling protocol called the GPRS Tunnel Protocol (GTP) is used to encapsulate multiprotocol data packets across the backbone [24]. GTP uses either TCP as a reliable transport mechanism required by X.25 data packets or UDP as a ‘best efforts’ base transport for IP packets.

The question then arises: how do we transfer IP or X.25 packets from the MS to the SGSN? This is performed by the *Subnetwork Dependent Convergence Protocol* (SNDCP), which is responsible for supporting compression techniques such as V.42 bis data compression and

TCP/IP header compression to improve channel efficiency [25]. In the event that other network layer protocols are required to be carried, the only change necessary would be to SNDCP itself. Details of how the protocols interact in the GPRS transmission plane [23] are shown in Figure 9.7.

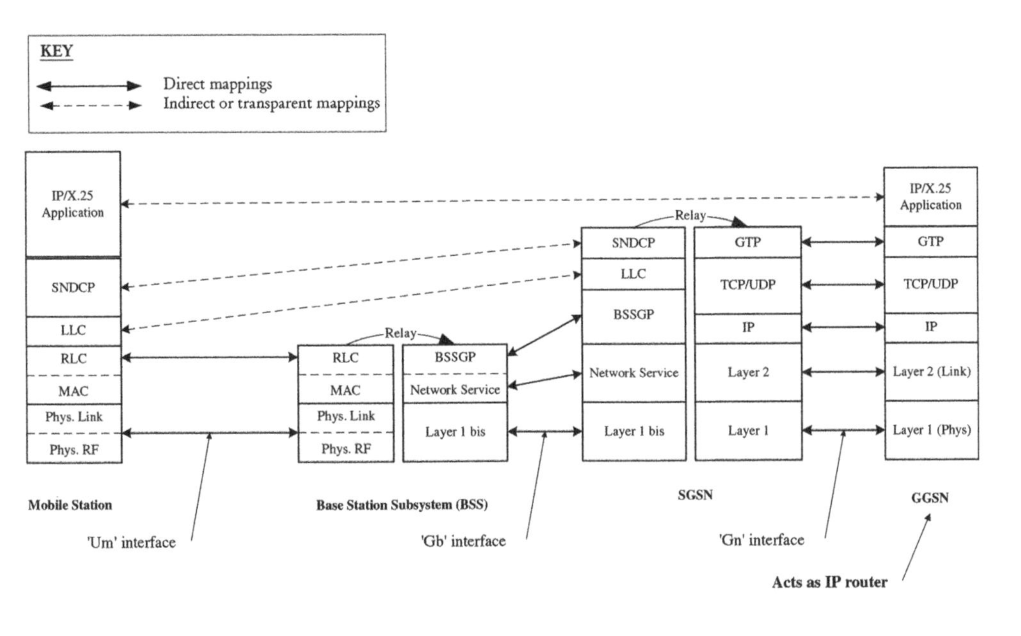

Figure 9.7 GPRS MS to network reference model.

Travelling down the protocol stack, packets become *frames* in the *Logical Link Control* (LLC) layer, which operates on the Um and Gb interfaces between the MS and the SGSN and is used for flow and sequencing control and ciphering, and can provide transmission error detection and recovery in *acknowledged mode*. In *unacknowledged mode*, it will simply flag up unrecoverable errors. LLC is also used to pass control data for Mobility Management (see Section 9.2.3.2 Mobility Management) and to forward SMS messages on behalf of the SMS protocol.

Error checks can also be performed at the next level down, at the *Radio Link Control/ Medium Access Control*[9] (RLC/MAC) combined layer, and these operate independently of the LLC, which has its own acknowledgement procedures [26].

The *RLC function* relays the frames from the LLC layer to the MAC layer, segmenting and re-assembling them into RLC *data blocks*, providing Backward Error Correction through selective retransmission of erroneous data blocks using a technique called Selective Repeat–Automatic Repeat Request (SR-ARQ). The Block Check Sequence (BCS) error check is carried out by the Physical Link Layer and not the RLC layer.

The principal *MAC function* is to control when multiple mobiles request access to the same BTS on the same channel or on different physical channels. It resolves contention for

[9] This MAC should not be confused with the layer-2 **Media** Access Control of the ISO model, which uses the unique MAC addresses burnt into networking equipment's interface cards for transporting frames.

resources, provides collision detection and recovery and allows a single MS to use several channels at once[10]. It would seem, then, that the MAC layer must monitor both uplinks and downlinks, but on the downlink side, it only needs to queue packets and schedule access attempts to the MS, as contention is not an issue in that direction. More importantly for this study, the MAC layer is also responsible for priority handling, which will be expanded upon in Section 9.3.7 QoS Profiles.

The *Physical Link* layer provides a channel between the MS and the BSS. In accomplishing this, it must detect physical link congestion, control transmitter power and MS power procedures such as DRX, and monitor radio-link quality to initiate cell re-selection processes for handoffs to take place. It is this layer that extends the TDMA structure to produce packet data channels (PDCH), interleaving four 'bursts' from four consecutive TDMA frames, and combines them into one *radio block* as shown in Figure 9.8. It includes Forward Error Correction (FEC) to detect errors in coding schemes and crucially synchronization to correct variances in propagation delay [27].

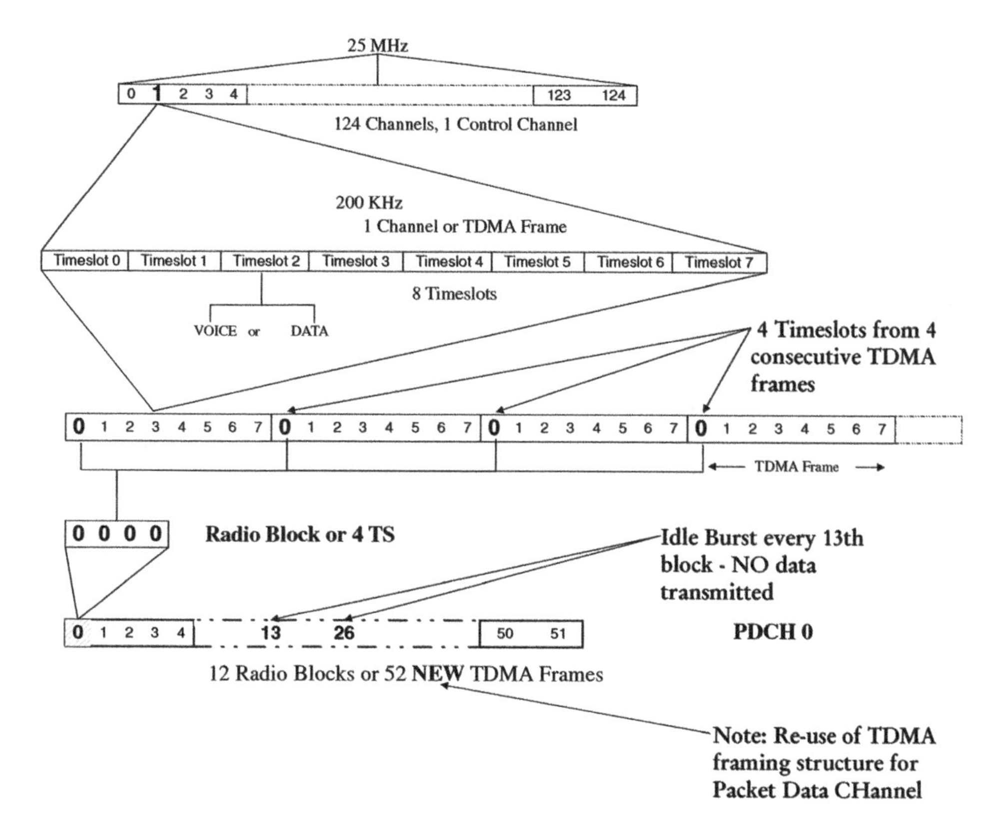

Figure 9.8 PDCH composition from TDMA frames.

[10] GPRS communications on the radio link are asymmetric (i.e. the number of channels used in one direction do not have to be the same in the other), and the two links, the uplink from MS to BSS and the downlink from BSS to MS, work independently of each other.

Finally, the *Physical RF* layer performs the same modulation of waveforms on the carrier frequencies in both directions as under the GSM architecture.

A GPRS timeslot is a PDCH. The basic transmission unit of a PDCH is a *radio block*, which is made up of four timeslots, known as '*bursts*', from four consecutive TDMA frames. Twelve radio blocks go to make up a single PDCH of 52 TDMA frames. The 13th radio block or '*idle burst*' carries no data.

Payloads[11] are light, headers are many (six, if we include the application header), and transmission times are long. The payload a radio block carries depends upon the message type and coding scheme, but less than 456 bits can be expected for each radio block with a transmission time of 20 ms per block. This equates to 240 ms for each PDCH sent. Of course not all the logical channels in a PDCH will be taken up with data. Just as we saw with GSM (Section 9.2.3.2 Mobility Management), there are a number of control channels that may be used. Most simply place the word *packet* in front of previous definitions. Thus:

- Packet Broadcast Control Channel (PBCCH) *downlink only*– GPRS system-specific information;
- Packet Random Access Channel (PRACH) *uplink only* – to initiate an uplink transfer;
- Packet Paging Channel (PPCH) *downlink only* – to page prior to a downlink transfer;
- Packet Access Grant Channel (PAGCH) *downlink only*– RR assignments are sent for both up and downlink transfers on this channel;
- Packet Data Transfer Channel (PDTCH) *up and downlink* – an MS can use multiple PDTCHs;
- Packet Associated Control Channel (PACCH) *up and downlink* – sends signalling information, acknowledgements related to an MS and its PDTCH.

9.3.3 Call Setup

To support the packet-switched nature of GPRS, each radio resource (RR) must only be assigned temporarily, which it does in a Temporary Block Flow (TBF). A BSC controls resources in both directions for data, voice, and control channels and will locate an MS by its MS address in the MAC block. Since all radio blocks come from the BSC, no concurrent access on any one PDCH can occur.

Uplinks, however, are contended for, and the BSC sends a 3-bit Uplink Status Flag (USF) to let each MS know which PDCH they have been assigned on the uplink for voice calls. For those mobiles wishing to initiate data transfers, the PRACH common channel arbitrates between them.

An example of a simple call setup is shown in Figure 9.9, in which requests are made, and resources are assigned using PRACH. If RACH from GSM phones is used, there are only two values for GPRS: one used to request limited resources and another to request two phase access. The second phase takes place using the PACCH as shown and describes the resource reservation more comprehensively.

[11] The amount of data carried within a packet compared with its header information.

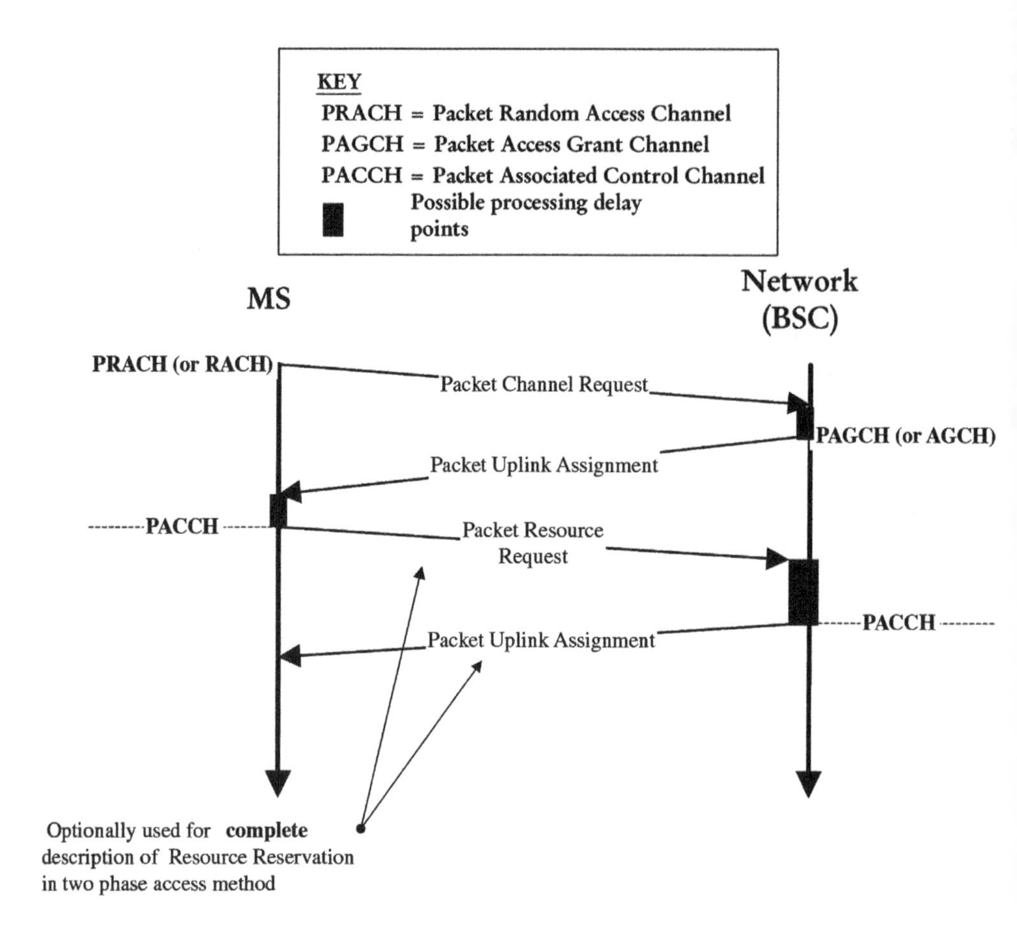

Figure 9.9 Call setup for uplink packet transfer.

9.3.4 The Air Interface

If we wish to maintain a reasonable level of performance over the GPRS network, the procedures for handoffs over the air or 'Um' interface must cater for both GSM circuit-switched voice calls and GPRS packet-switched data calls. A note from the 3GPP hints at one possible outcome of this attempt to converge voice and data:

> .... When a class A[12] mobile station is simultaneously involved in a circuit switched service and in a GPRS transfer...handover for the circuit switched service has precedence over GPRS network controlled cell re-selection. When a MS is performing Anonymous Access, cell re-selection implying a change in Routeing Area results in the MS returning to the GPRS MM IDLE state ... there might be cases where the network controlled re-selection would result in the Anonymous Access failing. [26]

[12] See Section 9.3.4.2 RR Operating Modes and Corresponding MM Modes for more information.

To understand the vagaries that an IP application may be subjected to in this new converged world, we must examine the following:

- Dual Transfer Mode (DTM);
- Radio Resource operating modes and Mobility Management states;
- Cell re-selection.

9.3.4.1 Dual Transfer Mode

This mode allows a MS to have an ongoing resource of one or more physical channels allocated by means of a Temporary Block Flow (TBF) in one direction while having a concurrent TBF established in the opposite direction allowing the MS to hold voice and data calls open at the same time. This is only available to Class A mobile phones that are DTM-enabled.

An important consideration from a performance perspective is that when moving to a new cell, the MS drops into a '*dedicated mode*' (i.e. voice-only mode) temporarily while system information is passed to it about the new cell. It then passes back into DTM, and the transfers can continue. While this may not effect file transfers, which have the ability to pause and continue, video playback and real-time video would suffer a degree of freeze frame or perhaps find that the audio stream continues ahead, or instead, of the video stream as currently often happens with web-based multimedia applications transmitted across slow links.

We should now look at the other modes available and the co-ordination between which modes the MS thinks it is in and the network's view of the same state

9.3.4.2 RR Operating Modes and Corresponding MM Modes

Under GSM, there were two RR modes: idle and dedicated. To this have been added two more: *packet idle mode* and *packet transfer mode*. Of the three classes of MS (A, B, and C), only Class A can support DTM. Some Class A mobiles and all of Classes B and C do not support DTM, and Class C mobiles are limited to being in either GSM or GPRS attached modes only.

A Class A GPRS MS supporting DTM can release the packet resources only and return to *dedicated mode* to continue with a voice call, for example, or release all RR resources and return to *idle/packet idle mode*. When in *idle/packet idle mode*, a Class A MS may initiate different voice RR or data packet resources simultaneously (always passing through *dedicated mode* to enter DTM[13]), while those MS in Classes B and C must leave both *packet idle* and *packet transfer modes* before going into *dedicated (voice) mode.*

This means that in a class B and C MS, the voice and data services are separate entities and may potentially offer a good chance of delivering service levels, whereas a MS that is capable of having both voice and data channels open will look to preserve the voice service over and above the data service. The three classes of operation and the transition between different states are shown in Figure 9.10 below.

[13] GPRS Data services are very much a supplementary service to the GSM voice service, as one would expect of an overlay network, and, as demonstrated above [26], the default control will reside with the GSM service.

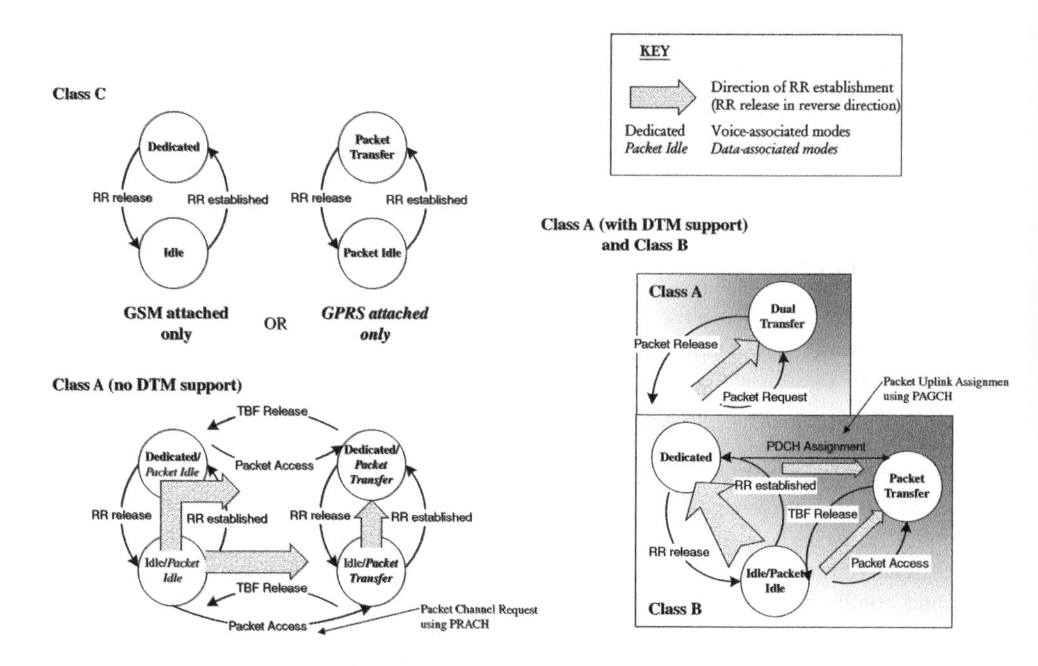

Figure 9.10 RR operating modes and state transitions.

The Network or NSS view of these states is represented in Table 9.1. Each voice-established state is protected by a timer for synchronization purposes, which runs in the MS as well as in the NSS. Packet transfers, however, have their own set of timers supplied by the RLC protocol.

9.3.4.3 Cell Re-selection

A *cell* is a geographic area served by a low-power transmitter providing enough power and frequencies (through frequency re-use[14] techniques) to support multiple subscribers. The location of the transmitter (BTS) is called a *cell site.*Traditionally, cells are represented by hexagons, but they can be any shape, e.g. rectangular, along a motorway.

Changing cells is governed by two factors: signal strength and channel quality (i.e. interference encountered). These have to be measured and the normal division of labour states that the MS shall monitor the downlink and the BSS the uplink. However, in the event that an MS is busy or forgets to make the measurements in either *packet idle* or *packet transfer mode*, the BSS (network) can force the suspension of normal cell re-selection and order the MS to take measurements before continuing. In the case cited above (Section 9.3.4 The Air Interface), where a failure of Anonymous Access was possible, the MS should stop

[14] Frequency re-use within a cell is essential where capacity needs are great. Using the free-space loss property of radio waves (i.e. signal weakens the further it propagates), smaller cells requiring a lower power can multiply the number of channels available by a so-called *N*-factor. Experience has shown $N = 7$ to be the optimum choice for low interference.

Table 9.1 RR operating modes and mobility management states

RR modes and corresponding MM states (MS without DTM)

RR BSS	Packet transfer mode	Measurement report reception	No state	No state
RR MS	Packet transfer mode	Packet idle mode		Packet idle mode
MM (NSS and MS)	Ready			Standby

RR modes and corresponding MM states (MS with DTM capability)

RR BSS RR MS	Dual transfer mode	Dedicated mode	Packet transfer mode	Measurement report reception	No state	Dedicated mode	No state
				CS idle and packet idle			CS idle and packet idle
GMM (NSS and MS)	Ready					Standby	

sending measurements and ignore cell change orders [26]. We are not told, however, how this particular contention should be resolved.

When measuring the received signal strength of frequencies on the downlink, these are taken on one of the packet common control channels (PCCCH) or on a broadcast control channel (BCCH), rather than voice or data channels, and are sent as RLC data blocks. The reason for choosing control channels is that signal strength relies on the transmitter power of the current or *serving cell*, and this power output is constant for PDCH frames that contain packet broadcast (PBCCH) or packet paging channels (PPCH). On all other channels, downlink power control could be used, which can fluctuate. A received level average (RLA) is calculated on the serving and neighbouring cells, and a choice is made.

When measuring channel quality on the downlink, checks on the interference levels are performed using the 13th *idle burst* of the TDMA frame (see Section 5 GPRS Lower Layer Protocols) or optionally on the timing channels (PTCCH) responsible for synchronization. The MS takes whichever of the idle burst or the timing channel has the lower signal strength and measures the interference. Again, a running average is maintained.

When in packet transfer mode, an MS looks at as many of the eight channels (or TS) as it can on an individual PDCH frame to take interference level readings and combines them with power, received signal strength, and signal variance into a channel quality report carried in a packet downlink acknowledgement message.

On the BSS side, things are much simpler in that it measures signal strength and channel quality on all uplink PDCHs (including packet transfers) whether active or not.

An impact on performance may be felt when an MS uses multiple timeslots (or *multislot*) and cannot take measurements. The BSS forces ‘periods of inactivity’ [26] to allow measurements to be taken and to identify its controlling BTS.

Another element of wireless networks that has an indirect impact on performance is *coding*, which is where we turn our attention to next.

9.3.5 Coding Schemes

Channel coding refers to the way in which data packets are compiled, coded and interwoven (or interleaved) on to the packet data transfer channel (PDTCH). There are four coding schemes[15] proposed for GPRS data packet channels [*Coding Schemes* (CS)-1 to -4]. Control channels except random access (PRACH) and timing control (PTCCH) all use CS-1. For a network to support GPRS, it need only be aware of CS-1, but GPRS mobiles need to support all four coding schemes.

In the assembling of data packets, all schemes add BCSs to detect errors and CS-1 to CS-3 employ a forward error correction (FEC) technique known as *convolutional coding*. CS-4 is a simpler scheme with no convolutional coding and therefore no error correction. The need for a faster, simpler scheme that reduced the amount of 'wasted' space (known as a *header*) in a packet that was not actual payload data over a network that had precious little bandwidth to offer in the first place must have been overwhelming. In many ways, this is analogous to UDP in IP data networks, except that the Internet could boast vastly reduced levels of physical errors with the introduction of fibre technology. The air interface has no such luxury.

The effect in performance terms, then, is a gain or loss in the amount of data that can be transmitted (or *data rate*) dependent on the scheme used. Table 9.2 details how the size of the data packet or radio block changes with the different schemes. Convolutional coding inflates and then deflates (or *punctures*) the size in carrying out its error correction. To many, this would seem surplus to requirements. All numbers signify data 'bits' (i.e. 1s or 0s).

Table 9.2 GPRS coding schemes and data rates

Coding scheme	Convolutional code rate	Uplink status flag (USF)	USF pre-coding	Radio block payload	Block check sequence (BCS)	Tail bits (padding)	Radio block after coding	Effective data rate
CS-1	Half rate	3	3	181	40	4	456	9.05 kbps
CS-2	Two-thirds	3	6	268	16	4	588	13.4 kbps
CS-3	Three-quarters	3	6	312	16	4	676	15.6 kbps
CS-4	Full	3	12	428	16	n/a	456	21.4 kbps

A glance at the final column is a stark reminder that the amount of radio resources on a single radio block available for passing any data traffic is *extremely limited.*

The enhanced version of GPRS called EGPRS, which is outside the scope of this study, is worth mentioning for two reasons. First it has nine modulation and coding schemes (MCS1 to MCS9), which propose different data rates from 8.8 kbps worst case up to 59.2 kbps best case. They are grouped into three families, and if the radio-link quality is poor, retransmissions can be directed over another MCS from the same family. While this may appear to provide some redundancy in frequency adaptation, it will do nothing but create havoc for an application that

[15] Enhanced GPRS or EGPRS has a total of nine modulation and coding schemes (MCS-1 to MSC9).

is now being asked to adapt to a new data rate! So-called *elastic* applications will cope, but real-time video and other *inelastic*or *non-adaptive* applications like their bit rates to be constant. Second, EGPRS manages to extend the data rates up to 59.2 kbps, but when one considers that, until fairly recently, it was thought that video-conferencing should be allocated a minimum of 384 kbps (or the equivalent of six ISDN channels), it does not bode well for adapting multimedia applications over wireless networks. As we will see, today's streaming applications and advanced compression techniques allow this to be dramatically reduced. However, the challenge still remains, and solutions need to be found. One such possible solution is the multislot capability.

9.3.6 Multislot Capability

Remember that a radio block consists of four identically numbered timeslots across four consecutive TDMA frames. In this way, one data channel (PDCH) can carry several data transfer channels (PDTCHs) for different mobile stations. A MS with multislot capability is able to use several data transfer channels (PDTCHs) across *multiple data channels*(PDCHs) to increase data rates. The multislot class determines the number of timeslots a MS can use.

There is now a ray of hope for mobile networks to be able to support multimedia applications, and this fact was taken into consideration when the laboratory testing was carried out. In effect, the conservative view was taken that up to two TS should only be considered, because rather like the older transmitters that were high-powered but had limited reach in terms of subscriber capacity, if one subscriber were to reserve a large number of timeslots, he or she might be the only one with access to the service. This clearly makes little economic or business sense.

Now that we have the possibility of something approaching a sufficiently high data rate (or *bandwidth*[16]), we can examine what other methods are available to add guarantees to that service.

9.3.7 QoS Profiles

At the time of writing, there is only one such method that is more laying down guidelines than guaranteeing service levels, which bears closer scrutiny.

Two of the deliverables of a GPRS network were to provide user differentiation based on QoS and volume-based charging. It is not difficult to see how these two are linked. Offering a premium service comes at a price, as we have seen with pay-per-view television channels. Equally, in the wireless environment, the 'quality' portion of QoS infers not necessarily a poor video picture quality, for example, but more the ability to receive video pictures at all.

To derive maximum benefit and therefore maximum revenues from QoS, it must have the capability to be applied on a per-user, per-application, and, ideally, per-application flow basis. To do this requires several changes to the entries in the information stored about the subscribers in the HLR database. It now needs to create *QoS profiles* for users and applications, which it does through the use of PDP contexts. Each PDP context is associated with a QoS profile, which is viewed as a single entity with multiple, independent values (or *classes*) for the transfer of data packets. The five *classes* are:

[16] Bandwidth is taken to mean data transmission rates in bits per second rather than spectrum frequencies.

- Precedence class – three priorities;
- Delay class – four classes;
- Reliability class – five classes;
- Peak throughput class – nine classes;
- Mean throughput class – 19 classes so far.

A QoS can be a combination of any number of the above, and it is not incumbent on an operator's network (PLMN) to provide anything but a limited subset of these profiles [27]. For the most part, the lowest common denominator of a best-efforts approach is assumed to spur the PLMNs on to better things.

As discussed in Section 9.3.3 Call Setup, an MS can make an optional packet resource request by selecting a RLC/MAC *radio priority level* and by indicating that it is sending data or control (signalling) information,during which time it can negotiate values for each of the above classes. There are four radio priority levels and an additional level for signalling. The network in the shape of the SGSN responds according to the available RR resources, determining radio access precedence (who goes first) and service precedence (who goes first in case of a network fire or congestion). In the case of SMS messages, the MS has no say in the matter. Once the negotiations are finished and PDP contexts modified, the PDP contexts are then activated and the application data flow and the fun can begin.

9.3.7.1 Precedence Class

The *precedence class* defines what actions, such as discarding packets, should be taken under congestion conditions and has three levels: high-, normal-, and low-priority.

9.3.7.2 Delay Class

The *delay class* determines the per-packet delay across the GPRS network, attempting to impose a level of performance, which takes account of numbers of subscribers in a cell at any one time as well as RR and network resources. There are four classes, and examples for packets of 128 bytes (or octets) are given [28] in Table 9.3:

Table 9.3 Delay class examples

Delay class	Class 1	Class 2	Class 3	Class 4
Mean delay	< 0.5 s	< 5 s	< 50 s	Best efforts
95% percentile	< 1.5 s	< 25 s	< 250 s	Best efforts

The thought of a best-efforts service, being the minimum requirement for PLMNs, does not fill one with a sense of overwhelming expectations.

9.3.7.3 Reliability Class

The *reliability class* looks at error rates linked to the probability of data loss, out-of-sequence, duplicate, or corrupted data, endeavouring to satisfy the requirements of the network layer protocols such as TCP and UDP in combination with GTP. To achieve this, GTP, LLC and RLC layers (Section 9.3.2 GPRS Lower Layer Protocols refers) must support the classes [27] listed in Table 9.4.

A noticeable omission from Table 9.4 is for real-time traffic that does *not* tolerate data loss. However, the note acknowledges the need for a combination profile of reliability, delay, and throughput. The more complex notions of one-way delay and consecutive packet loss that would be needed to support a VoIP application, for example, have not yet made an entrance into these standards.

9.3.7.4 Peak Throughput Class

The *peak throughput class* can be thought of in a similar way to the peak cell rate (PCR) definitions for ATM or the 'burst rate' or extended information rate (EIR) of Frame Relay. Throughput is measured from the R to the Gi reference points (the shaded area of Figure 9.6) from a MS to the point at which it enters an IP packet data network. In IP terms, this is considered to be one routeing hop. The peak throughput is independent of the delay class and is measured in bytes per second, as shown in Table 9.5. There are no guarantees that this peak rate will be achieved or sustained for any length of time: it is merely a limiting boundary beyond which the subscriber would need to negotiate further, if an application demanded it, and presumably pay more for the privilege.

9.3.7.5 Mean Throughput Class

The *mean throughput class*, however, does maintain an average data rate that can be expected during the lifetime of an active PDP context, which for now we will take to be an application flow. It is envisaged that a negotiated mean data rate would support a flat rate charging scheme.

The reference points R and Gi remain the same for the measurements, but note that the data rates are now measured (Table 9.6) in bytes *per hour*. An approximate equivalent data rate is provided to emphasize the conceptually low data rates expected of the GPRS service.

The standardization of QoS carries on, and much work is still to be done to align QoS in GPRS with the next generation of mobile technology (3G or UMTS). Once complete, this will enable roaming between the two networks with the same QoS applied, but we still need to know if the currently available GPRS networks will support some, or any, of the multimedia applications. On the evidence of the available data rates, it would seem that their chances are fairly slim.

Table 9.5 shows other possible data rates that are being used in developments designed to increase the peak throughput per timeslot for both high-speed circuit-switched data (HSCSD[17]) and GPRS, which are currently limited to 64 kbps and 160 kbps, respectively. One of these developments, enhanced data rate for global evolution (EDGE) is promising to deliver 384 kbps with 'full mobility', spawning two new services in the process: enhanced circuit-switched data (ECSD) and EGPRS. The phrase 'full mobility' seems curious, until

[17] HSCSD is a circuit-switched rather than packet-switched service that shares the ability with GPRS to allocate more than one timeslot per user.

Table 9.4 Reliability classes[a]

Reliability class	GTP mode	LLC frame mode	LLC data protection	RLC block mode	Traffic type
1	Acknowledged	Acknowledged	Protected	Acknowledged	Non-real-time traffic, error-sensitive, no data loss possible
2	Unacknowledged	Acknowledged	Protected	Acknowledged	Non-real-time traffic, error-sensitive, infrequent data loss supported
3	Unacknowledged	Unacknowledged	Protected	Acknowledged	Non-real-time traffic, error-sensitive, data loss supported (e.g. signalling and SMS)
4	Unacknowledged	Unacknowledged	Protected	Unacknowledged	Real-time traffic error-sensitive, data loss supported
5	Unacknowledged	Unacknowledged	Unprotected	Unacknowledged	Real-time traffic error non-sensitive, data loss supported

[a] Note: For real-time traffic, the QoS profile also requires appropriate settings for delay and throughput.

Table 9.5 Peak throughput classes

Peak throughput class	Peak throughput (bps)	Equivalent data rate (kbps)
1	1000	8
2	2000	16
3	4000	32
4	8000	64
5	16,000	128
6	32,000	256
7	64,000	512
8	128,000	1024 (or 1 Mbps)
9	256,000	2048 (or 2 Mbps)

one considers that the wideband code division multiple access (W-CDMA[18]) technology that permits a maximum speed of two megabits per second remained available only in static mode [29] as late as February 2001! Basic assumptions, it would seem, cannot be taken for granted, and trials such as that on the Isle of Man will be the proving ground for this technology.

Table 9.6 Mean throughput classes

Mean throughput class	Mean throughput (bytes per hour)	Equivalent data rate
1	100	0.22 bps
2	200	0.44 bps
3	500	1.11 bps
4	1000	2.2 bps
5	2000	4.4 bps
6	5000	11.1 bps
7	10,000	22 bps
8	20,000	44 bps
9	50,000	111 bps
10	100,000	0.22 kbps
11	200,000	0.44 kbps
12	500,000	1.11 kbps
13	1,000,000	2.2 kbps
14	2,000,000	4.4 kbps
15	5,000,000	11.1 kbps
16	10,000,000	22 kbps
17	20,000,000	44 kbps
18	50,000,000	111 kbps
31	Best effort	

[18] W-CDMA forms the basis of UMTS and its next step evolution IMT-2000. US flavours include 1XRTT (144 kbps) as a stepping stone to their own CDMA-2000, higher data rate (HDR) at 2.5 Mbps and 1 EXTREME at 5 Mbps.

The upper bounds of Table 9.5 are more applicable, then, to the next generation of mobile networks, i.e. the 3G or Universal Mobile Telecommunications System (UMTS).

9.4 UMTS (or 3G) Networks

It is not intended to go into any great detail with 3G networks, as they are continually evolving, but key parts of the standards are in place. We will concentrate on aspects of the architecture and mandated standards that affect performance levels of services and applications.

We will examine what has changed in terms of architectures and traffic classes to see how QoS and application requirements are being met and why those classes, under their GPRS definitions, were not sufficiently comprehensive to cater for applications coming on to the market, such as audio- and video-streaming.

We also look ahead to the review of one particular mandated standard that is not only enabling QoS but is altering the face of IP data networks through its rightful insistence on end-to-end QoS performance guarantees. I refer to the use of IPv6 as a pre-requisite in that heart of a UMTS network that deals with such elements as call setup, subscriber profiles and mobility, which is known as the IP Multimedia Subsystem (IMS) and bridges between the wireless access networks and other IP data packet networks.

9.4.1 UMTS (or 3G) Architecture

The UMTS network is not a new network but an evolution from GSM and GPRS networks, standardized by the 3GPP organization, to maximize the benefits of the existing networks. The longer-term solution of an 'all-IP network' was handed to another organization in early 2000, called the Mobile Wireless Internet Forum (MWIF).

That evolution will take place over a number of years, and to accommodate this, the different stages have been given release numbers. Unfortunately, the 3GPP fell into the same trap as a large desktop software organization by deciding to name their versions after the years they were planned for release. Consequently, the version being deployed now in 2001 is release R99, while R00 may appear sometime in 2003.

We will focus on the R00 architecture, shown in a simplified form below (Figure 9.11), for two reasons. First, the R99 version only supports ATM as its transport protocol and W-CDMA as its radio access technology. Second, as discussed previously (Section 9.3.7.3 Reliability Class), real-time applications that cannot tolerate data loss were left undefined in R99. If we are to study real-time applications and their performance levels, we must relate the study to a level of QoS that recognizes those types of applications and it must be in the context of an IP transport layer. UMTS R00 defines two RAN methods: the GPRS/Edge radio access network (GERAN), as previously discussed, and a UMTS terrestrial radio access network (UTRAN) using wideband CDMA.

Since it is an evolution, the architecture comprises three sub-networks or domains to preserve existing network deployments: circuit-switched (CS) and packet-switched (PS) domains and a RAN domain to include both GERAN from the GPRS model and UTRAN from the new UMTS model. Notice that the two CS and PS domains still exist, the CS domain providing voice services linked to the PSTN and ISDN networks, while the PS domain supplies data services and Internet connectivity.

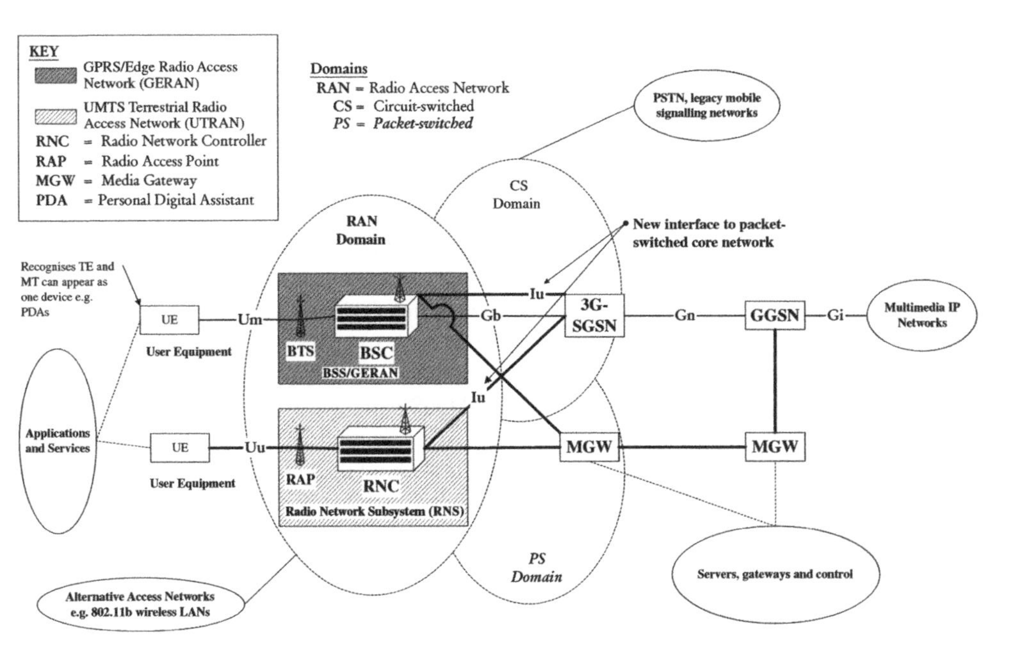

Figure 9.11 UMTS R00 architecture.

The main objective of R00 is to use the same UMTS core network for different RANs to connect to, including other wireless technologies, such as Wireless LANs (WLAN 802.11b[19]) that have made an appearance this year with the relaxing of government rules on transmitting wireless spectra in public places and Broadband RANs.

9.4.2 QoS Service Architecture

Under the UMTS architecture, a good deal of contemplation over service definition has taken place. The result is a layered bearer[20] QoS service architecture, which appears overly complex for the task in hand. However, this is bound to happen when defining end-to-end services across different media, as the services promised by one transport medium (TCP/IP, ATM, Frame Relay, X.25) must be translated into services defined at other transport levels (radio access at RLC/MAC level, FDMA/TDMA multiplexing of PDCH blocks and physical layer). The end-to-end service is broken down into three components:

[19] This IEEE Ethernet standard comes in 2-Mbps and 11-Mbps versions. Being a shared medium, the actual throughput is around 840 kbps for the latter version. Proposed extensions to 802.11b will see speeds of 22 Mbps at 2.4 GHz, whereas a proposed Ethernet 802.11a standard promises 54 Mbps at 5.7 GHz.

[20] A 'bearer' in voice networks is synonymous with a 'trunk' to mean a physical cable or link carrying voice traffic. This is distinct from the 'carrier', which is the electrical signal of constant frequency whose modifications are the voice or data information carried.

- the local bearer service at the mobile station, now user equipment (UE) or TE/MT if two separate devices are used, e.g. a laptop with a mobile phone;
- UMTS bearer service – bridges the gap between core network bearer service and the radio access bearer service and takes account of user profiles and mobility;
- external local bearer service – connects the core network service to an external network service.

The full UMTS QoS service architecture is shown in Figure 9.12. Two aspects about this plan will seem curious. First, the backbone network service mentions only layers 1 and 2 of the ISO seven-layer model. This is partly historical in that GSM, GPRS, and UMTS R99 have relied on ATM, which is a layer 2 or link-layer protocol, for transporting data packets using ATM adaptation layer (AAL) 5 as well as AAL2 for voice if multiplexing voice and data are required. ATM has clearly defined and recognized QoS parameters in place, whereas TCP/IP and the IETF are still putting flesh on those bones (see Section 10.7).

Second, no end-to-end service would be complete without the knowledge and participation of the client, in this case the mobile device (renamed variously as TE/MT and UE to take account of product developments). In developing an all-IP network, now under the aegis of the MWIF, this would have to be a prerequisite.

It is expected that these bearer services will have their own QoS attributes, some of which will be shared. The essential changes in defining the UMTS QoS service architecture were the re-defining of traffic classes to include different types of real-time applications and several additional attributes linked to those traffic classes.

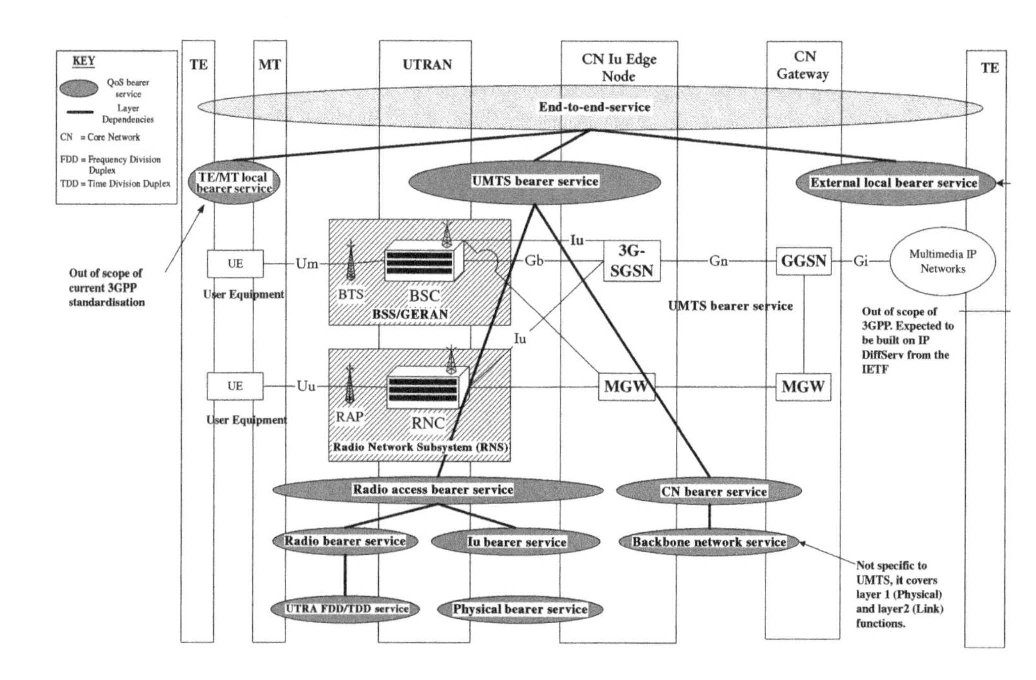

Figure 9.12 UMTS QoS service architecture.

9.4.2.1 QoS Traffic Classes

These fall into four categories that are defined by their real-time needs [29] as follows: (Table 9.7).

Table 9.7 UMTS QoS traffic profiles

Traffic class	Conversational	Streaming	Interactive	Background
Fundamental characteristics	Preserve time relation (variation) between information entities of the stream Conversational pattern (stringent and low delay)	Preserve time relation (variation) between information entities of the stream	Request response pattern and preserve payload content	Destination is not expecting content within a certain time Preserve payload content
Application example	Voice	Streaming video	Web browsing	Background download of e-mails

Conversational and streaming classes are both designed to meet the needs of real-time applications, whereas interactive and background classes refer to those only needing a best-efforts response. Each class will make use of different attributes according to its needs. For example, the transfer handling priority is of no consequence to conversational or streaming real-time applications, whereas the transfer delay attribute is critical to them. Table 9.8 lists the attributes by traffic class, some of which have already been discussed in the GPRS definitions (Section 9.3.7 QoS Profiles).

Table 9.8 UMTS bearer attributes by traffic class

Traffic class	Conversational class	Streaming class	Interactive class	Background class
Maximum bit rate	✓	✓	✓	✓
Delivery order	✓	✓	✓	✓
Maximum SDU size	✓	✓	✓	✓
SDU format information	✓	✓		
Delivery of erroneous SDUs	✓	✓	✓	✓
Residual bit error rate	✓	✓	✓	✓
SDU error ratio	✓	✓	✓	✓
Transfer delay	✓	✓		
Guaranteed bit rate	✓	✓		
Traffic handling priority		✓		
Allocation/retention priority	✓	✓	✓	✓

The attributes' functions can be described briefly as follows.

- maximum bit rate – equivalent to GPRS peak throughput;
- guaranteed bit rate – equivalent to GPRS mean throughput;
- delivery order – in-sequence delivery required or not;
- maximum SDU[21] size;
- SDU format information – a list of possible SDU sizes to optimize scheduling[22] over the radio interface;
- SDU error ratio – fraction of lost or errored SDUs;
- transfer delay – maximum delay of the 95th percentile distribution of delay among all SDUs through the duration of a call, i.e. 95% of all data packets must be below the maximum transfer delay value;
- traffic handling priority – equivalent to GPRS precedence class;
- allocation/retention priority – relative importance of resource retention and allocation between bearer services.

Through these extensions to the attributes and traffic classes, the needs of real-time applications appear to have been catered for. Although not specifically stated, one can assume that the conversation class includes video-conferencing as it involves two-way, asymmetric packet transfer. What is not clear at the time of writing is whether both one-way delay and round-trip delay measurement are included. The former would most certainly be needed for VoIP applications.

9.5 Implications of a 3GPP Mandated Standard

QoS in IP data networks was considered by many network managers, working in enterprises, carriers, or ISPs, to be an unnecessary evil that could be put off if one did not envisage using VoIP or promoting it as a service. An equally unnecessary distraction was that of IPv6, which, through some clever sidestepping manœuvres such as classless inter-domain routeing (CIDR) on the WAN side and network address translation (NAT) where the WAN meets the LAN, was largely ignored, whereas IPv4 enjoyed a new lease of life. That is all about to change, as IPv6 is now a mandatory element of the next generation of wireless architectures.

With this element now centre stage, it is worth looking at how this QoS, once defined, will be carried in an IPv6 network and how requested service levels of performance will be preserved when carried across the packet network both within and outside the UMTS infrastructure. The next chapter examines the mechanisms that will assist in guaranteeing QoS when moving from fixed to wireless networks.

[21] A service data unit (SDU) is a protocol data unit (PDU) with a defined packet data protocol (PDP) context.
[22] 'Scheduling' can be thought of as the radio equivalent of 'queuing' data on IP routers.

10

Moving to Wireless

In trying to guarantee QoS for any application, one of the major considerations is congestion management [30]. In a traffic jam, nothing gets through: so it is with IP data networks.

In this chapter, we consider fixed IP networks and why their original mechanisms for dealing with congestion needed a radical overhaul to support the real-time, multimedia applications of today. First of all, we will need to address the fixed network legacy, namely

- QoS roots – to determine why we need QoS at all;
- effects of router scheduling and queuing – what happens to a packet when it enters and leaves a network;
- TCP operations – its pressures and efficiency gains;
- UDP operations – particularly when in the company of TCP traffic;
- RTP operations – the need for timing and sequencing.

to understand the behaviour and so-called 'bursty' nature of any IP data network. Second, we will move on to discuss techniques, some new and some not so new, that will contribute to the successful maintenance of quality of service parameters throughout a data packet's journey between fixed and wireless networks:

- Internet Protocol version six (IPv6);
- IntServ and DiffServ approaches to QoS;
- Multiprotocol label switching (MPLS);
- Effects of compression standards for multimedia applications.

If mobile devices are to become a gateway or 'portal' to be able to access such applications, a consistent method of getting packet A to location B in a regular and timely fashion is entirely necessary. If they are to become the portal to all applications, where voice calls can continue uninterrupted in the simultaneous presence of video replays and web browsing in what can be termed a 'multiservice network', mechanisms must exist to allow the peaceful co-existence of time-dependent, prioritized, and casual traffic without the presence of one being to the detriment of either of the other two. In much the same way, traffic schemes on our motorways see the emergency services using the hard shoulder, whereas bus and taxi lanes (or multioccupancy vehicle lanes on US expressways) are an indication of environmental priority over other casual users whose lives do not depend on arriving first at their destination, even though some may drive as if it does!

10.1 QoS Roots

'Why do we need QoS at all?' is an all too familiar cry that is answered in this section, which explains how the need for QoS arose, why contention or congestion management lies at the very heart of QoS, and why router queues are a necessary feature not only to alleviate but also to control congestion in support of QoS.

The IP network, as originally conceived, was designed to be a 'dumb' network with intelligent peripherals. Applications were conversations that took place between a server and its clients, and routers simply forwarded packets between one subnetwork and another, using their shared knowledge of all destination networks to complete its main task of finding a route from A to B.

There were no hard and fast guarantees that traffic would arrive at its destination, and any physical break (or *outage*) in the route would result in lost traffic, relying either on manual intervention to restore the connection by re-routeing it or on routeing protocols to find an alternative route. These outages now last a matter of 20 s with modern routeing protocols such as Open Shortest Path First (OSPF), but more typically would last several minutes while the new route was propagated to all routers (known as *route convergence*) or, in the case of physical breaks, days to find the break and repair it. This no-guarantees approach was aptly named a *best efforts* service and is still the basis of most private and public IP networks.

The drive to cut costs enticed businesses to look to the Internet as a viable means of transporting what were then non-mission-critical applications across a pseudo-private network overlaid on to the public Internet, known as a virtual private network (VPN). As web browsing became the low-bandwidth way of viewing applications, allowing remote access to mission-critical applications, guarantees about the security and reliability of that traffic came under the spotlight.

QoS is nothing new: it first appeared in the OSI model in the 1980s, when four different categories were identified:

- bandwidth or capacity in bits per second
- latency (or delay)
- jitter or variations in delay
- traffic loss

The first is required simply to support an application's throughput requirements over a shared medium. It has been the position of many network operators that simply 'throwing more bandwidth' at a problem will cure any problems to do with application response times and latency or delay. This shows a deep misunderstanding of the nature of TCP and the way in which it was designed to operate, which is addressed below in Section 10.3 TCP Operations.

The second normally describes the round trip time (RTT) between a client request and the server supplying a response, which includes all processing times at the server and intervening network nodes, such as switches and routers. For a fuller explanation of RTT, see Section 5.2 TCP Flows. Increasingly, with VoIP applications, one-way delay is a necessary measurement to ensure toll-grade voice quality in both directions. In order to support voice and video applications, latency is crucial. Packets that arrive too late cannot be used and are dropped, and those that are too early are discarded if there is insufficient memory or *buffer* capacity in the client or node controlling the voice or video session. Either way, discontinuous transmission occurs, and frustration sets in on the part of the receiver.

The third category, jitter, causes problems for multimedia applications in that it complicates the playback process. This is most commonly seen as packets wait in router queues for their turn to exit the node on its outbound interface.

Finally, traffic loss is devastating, as voice and video applications have no hope at all of retransmitting packets in a time frame that passes unnoticed by the listener or viewer. Moreover, compression techniques are such now that the probability of losing large pieces of conversation or video playback is very real (see Section 10.10 Compression Standards for Multimedia Applications).

There is, therefore, a very real need to manage the last three, which can be collectively brought together under the banner of *congestion.*The response to that need is twofold: the one reflects what will come to be regarded as the old, *best-efforts* way of working, the other the new, managed QoS way that takes responsibility for the traffic it handles. As Huston has it:

> Where demand exceeds capacity, some form of demand management is necessary: Either the end-systems work adaptively to establish a point of equilibrium of demand (the best effort response) or there is an imposed resource management policy that selectively delivers resources to particular network users (a managed QoS response). [31]

Regardless of the QoS techniques discussed later in this chapter, the fact remains that an essential part of congestion management lies with the router's abilities to manage its queues both in a responsive manner and in step with whatever external rules or policies are being sent to it.

10.2 Router Scheduling and Queuing

It is not intended to give a full account[1] of queuing and scheduling mechanisms, but merely to state what they are, what problems they try to address and why they are important when trying to guarantee a service level. We will consider the following:

- first in–first out queuing (FIFO)
- priority queuing
- class-based queuing (CBQ) and weighted round robin (WRR)
- weighted fair queuing (WFQ)
- packet discard, similar to random early detection (RED) mechanism in TCP

Once these have been considered, it will be seen which mechanisms are more suited to multimedia applications, which mechanisms operate more effectively over low-bandwidth links, which mechanisms are more efficient across a high-speed core, and which application types benefit from individual queuing mechanisms or from combinations of more than one mechanism. From the number of mechanisms involved, it is clear that one size certainly did not fit all.

For now, it important to note that routers are *stateless* devices in a largely *connectionless*[2] network, i.e. they have no concept of the current state of the network in their

[1] A full account of queuing techniques currently available can be found in [31]. For a comprehensive theoretical perspective on queuing and how it relates to applications and network switching environments, see Kleinrock [32].

[2] Whereas *connection-orientated networks* rely on reusable labels, which have local significance only, i.e. outages can be kept local and re-routeing achieved locally without the participation and knowledge of all devices.

immediate vicinity, but rely on global addressing states stored in their routeing tables. Consequently, when an interruption to a route occurs, the changes must propagate to all routers before the network is stable enough to begin forwarding packets again. Additionally, routers have no flow control or re-transmission mechanisms, leaving that to the sending device and its TCP. Figure 10.1 shows at a basic level what happens to a packet entering (at the ingress) and leaving (at the egress) a typical data network. Note the intense activity around swapping out different link layer (layer 2) headers to be able to forward a packet on to the next network segment. Adding the complexity of assimilating different QoS parameters between different LAN and WAN topologies increases the workload on the queues.

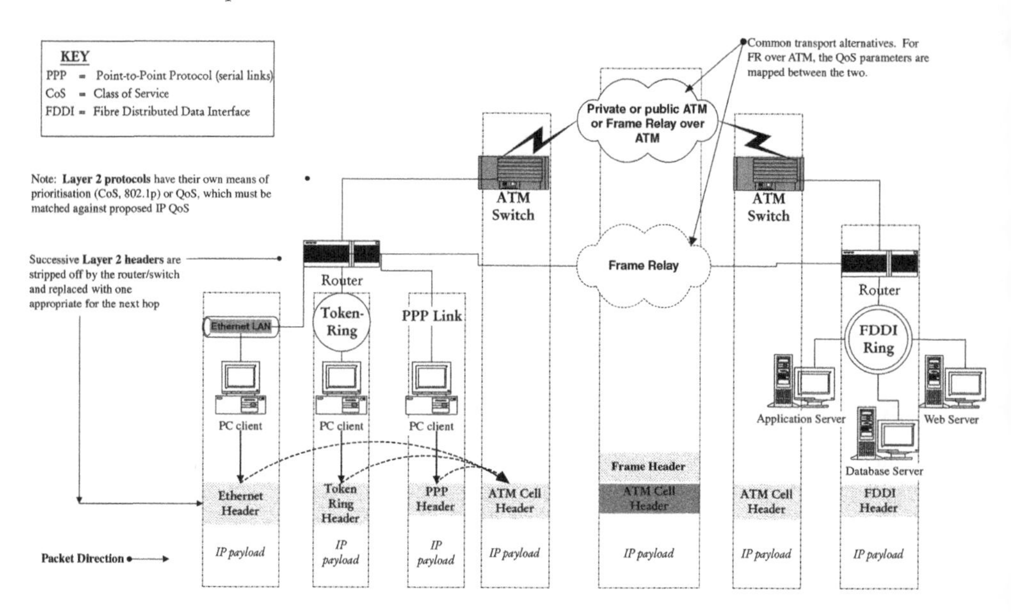

Figure 10.1 Journey of an IP packet.

10.2.1 FIFO Queuing

This is the base level offering of most routers and is simple and cheap to implement. Those packets arriving first in the queue are first out of it. The holding nature of these queues does not lend itself to applications with stringent latency requirements, and in congested periods, these queues exacerbate the problem by discarding all traffic once the queue is full. This in turn causes re-transmissions on a huge scale, only serving to magnify the problem.

This type of queuing is suitable only for best-efforts traffic and has formed the bedrock of the Internet for most of its existence.

10.2.2 Priority Queuing

Prioritizing certain types of traffic into high, medium, and low categories has been possible for some time by filtering on source and destination addresses or on application port number.

In this way, critical applications with known port numbers[3] could be given preferential treatment. On its own, this was doomed to fail because, while internal departments jockeyed to position their favourite application for priority treatment, the company e-mail system would grind to a halt under the watchful gaze of their CEO.

Unfortunately, this mechanism ignores the needs of the low- and even medium-priority traffic, removing the ability to signal back to their clients that they may be in for a long wait. TCP's response to this kind of congestion is covered in Section 10.3 TCP Operations. When all the seats on the bus are taken, it would seem like an eternity for an end user of a medium-priority application to wait for the next bus, but for the user of a low-priority application, he or she can only watch as bus after bus passes by packed to the gunnels with commuters escaping another day of tube strikes.

This type of queuing does not meet the need to be able to support all types of application fairly and equitably.

10.2.3 Class-based Queuing and Weighted Round-robin

Born out of the need to provide equity to all forms of traffic without starving any of them, class-based queuing assigns traffic to classes that can be prioritized as above in priority queuing. It allows for each queue to be defined by the number of bytes that are drained from it, which permits leeway in the rate at which packets are accepted based on an applications optimal packet or PDU size. Packets are drawn from each queue in strict rotation in a *round-robin*fashion, with *weighting*being provided by the size in bytes of a queue's service window. The effect of this is that there exists a service rotation interval or delay while other queues are being serviced.

An advantage of this system of queuing is that one can alter the treatment of an application by moving it from one class to another in short order. Another important feature is the freeing up of spare bandwidth for use by other applications, which is more efficient and fits neatly with the TCP mode of operation.

The main pitfall of this approach is that in mixing different application types, PDU sizes will vary greatly, and this runs the risk of those with larger packet sizes of 1500 bytes for bulk data transfer, for example, taking the lion's share of the bandwidth over an audio or video streaming application, which might average only 150 bytes. Interactive applications may be only 60 bytes long. One way of countering this unfairness would be to increase the service windows to accept more packets of all classes and thus even out the ratios somewhat, but this increases the service rotation cycle to the point that unacceptable delays are introduced as a result of applications waiting to be serviced.

Particularly hard hit by this would be low-speed links (extremely relevant when wishing to transfer packets to a wireless network) and when trying to pass applications bounded by low latency and jitter requirements.

However, on high-speed links of 155 Mbps and upwards, where transmission times are far greater than the service rotation cycle time, one can see that with a small number of queues,

[3] Applications have the equivalent of telephone numbers, where the area code is the application *port number*, and the remaining numbers are the *session number*. Servers traditionally have to 'listen' for application client requests on a known port number, while the client typically will generate its own random port number to which the server responds. Once the application flow begins, sequential numbering is also applied to groups of packets in the flow.

the levels of jitter and latency would be vastly reduced. This positions CBQ as a fair way of allocating bandwidth to a mixture of application types in the core of large networks.

10.2.4 Weighted Fair Queuing

In response to this seemingly unfair advantage gained by applications using larger packet sizes, a new mechanism was developed that did not operate a strict round-robin selection process but instead serviced higher-priority queues more frequently than lower-priority queues. Traffic with a heavier weighting would thus be serviced more regularly and their packets interleaved with the outgoing best-efforts traffic.

This approach focuses very much more on the *scheduler* responsible for placing incoming packets into the correct queues, as updates of current queue status must be taken into account before placing packets into queues and prior to their departure. In order to effectively ring-fence a service class and protect it from the demands of another a good deal of sampling of the queues is required to find out their current state. The processing needed to accomplish this at the core of a high-speed network, where updates would need to happen every few micro-seconds, is considerable, and not unsurprisingly, therefore, WFQ is deemed more suitable for low-speed links as an edge-control mechanism.

The ability to segregate individual application flows and interleave, for example, voice and video streams ensuring each receives its fair share of bandwidth to avoid latency and unnecessary packet drops puts it firmly in contention when contemplating running a multiservice network out to a wireless edge. WFQ does provide a minimum level of resource allocation to a service class independent of the activity levels in other classes [31], but can do nothing to regulate the jitter induced by the queuing. This requires additional admission control mechanisms to complement it.

One can conclude that queuing techniques are an important part of the jigsaw to provide effective service level performance control. A combination of CBQ for the high-speed core of a network and WFQ for low-speed links applied at the network edge to monitor application flows goes some way to creating an environment in which many application types with different latency and jitter requirements can co-exist. However, this is not enough to support multiservice networks. These techniques must be complemented by a common method for packet classification across multiple network topologies, admission controls both to regulate traffic entering a network and to mark discard eligibility preferences at an edge or at an end-user device, packet discard control once packets have entered the network, forwarding mechanisms that are responsive to local not global conditions, and a means of updating path information 'on the fly'. We continue by looking at the packet discard function.

10.2.5 Packet Discard

An essential part of congestion management is the last-resort measure of dropping packets. Queues will fill up, poorly written applications will misbehave, and resources particularly on slow links will be stretched.

> It is more stable to degrade incrementally than to wait until the buffer resources degrade to complete exhaustion in a single catastrophic collapse. [31]

Many network managers would probably admit readily to having experienced both such situations above, but this actually refers to a mechanism employed by TCP, called RED, which believes that it is better to drop packets when a threshold is breached rather than to wait for the queue to become saturated, by which time, the device has either dropped all packets or locked up completely.

The second function of the *scheduler*is to control the timing and selection of packets to be discarded. There are almost as many approaches to this as there are to queuing, but essentially, they fall into two camps: drops controlled by the routers and those controlled by the end device or by the device at the edge of a network responsible for admission control. The router-controlled drops very much dictate when and how, without reference to the application or the needs of its end-user, whereas those controlled by the edge or end device have the advantage of declaring what is essential and non-essential from the outset of the application flow. Examples of the latter are the discard eligible (DE) bit in Frame Relay and the cell loss priority (CLP) bit in ATM, which set the drop preference in packets prior to despatch. The IP header has no such bit to set a drop preference at present.

A router's basic scheduling, queuing and congestion-avoidance functions are shown in Figure 10.2, are followed by a summary (Figure 10.3) of the different queuing mechanisms that have been discussed above.

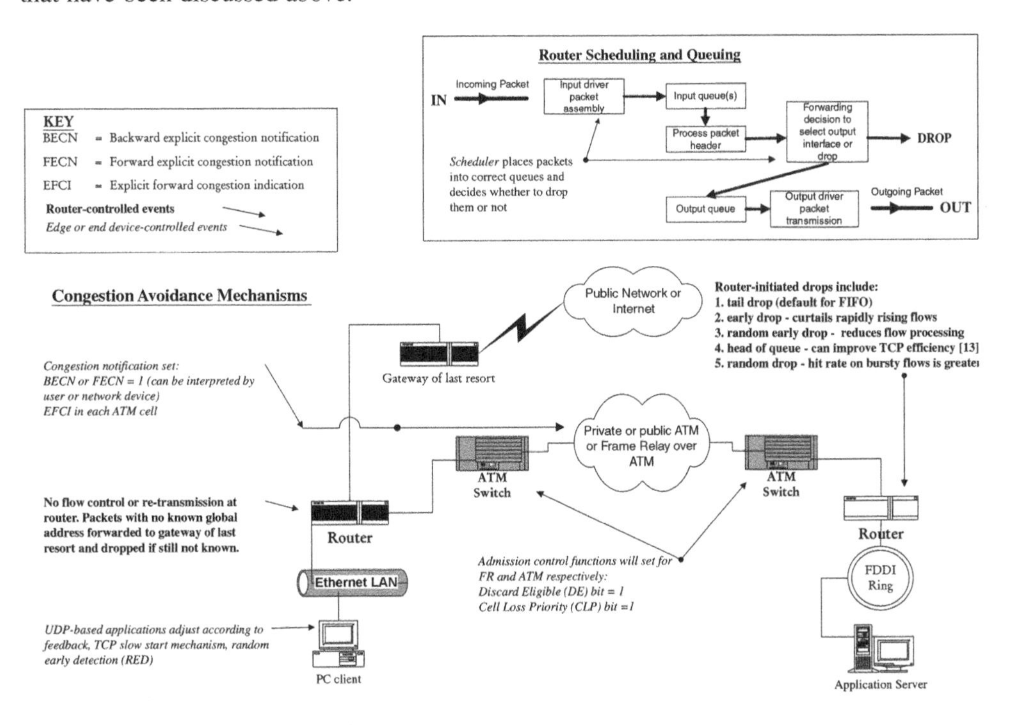

Figure 10.2 Scheduling and congestion management.

The intention is not to detail how congestion management techniques evolved in Layer 2 technologies such as X.25, Frame Relay, and ATM, as Black [33] provides a very good guide in that area. Suffice to say that these connection-orientated protocols satisfy many of the requirements of a QoS-capable network, including backward and forward error congestion

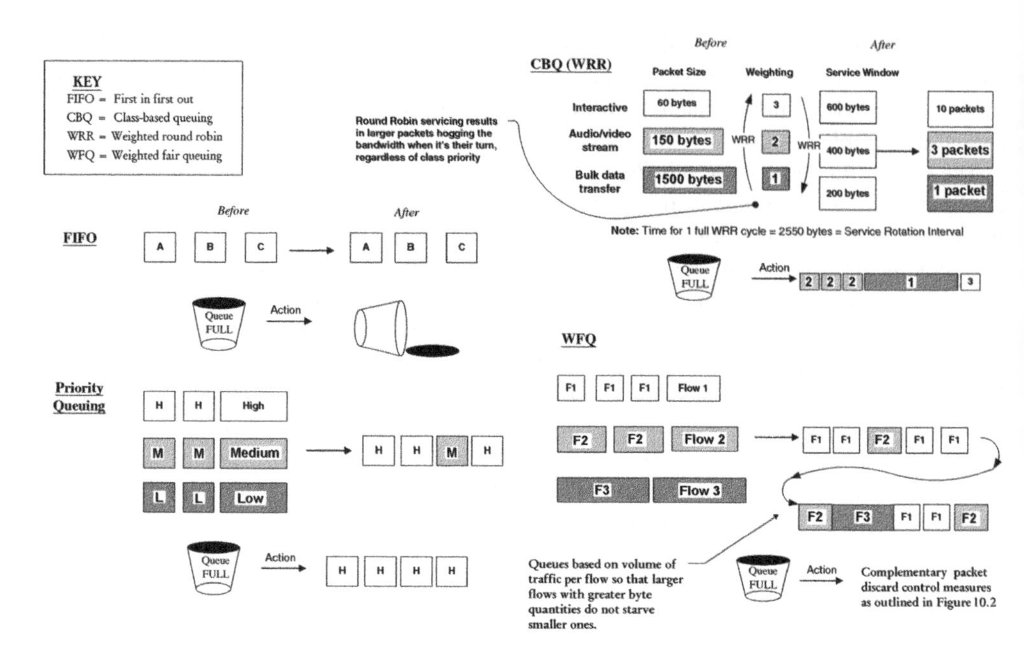

Figure 10.3 Router queuing techniques.

notification, path discovery, link-by-link route recovery and QoS definitions for controlled levels of latency and jitter. What they do not have is the end-to-end coverage that IP has.

Much of the work on queuing above and on QoS components below (see Section 10.7 QoS Approaches in Fixed Networks) revolves around introducing the controlled elements of connection-based switched networks, where negotiation of call setup parameters is commonplace, into the semi-controlled chaos that is the world of connectionless IP routed networks. One of the few perceived control elements in that IP world is TCP, to which we now turn our attention to understand its benefits and disadvantages in moving from a fixed to a wireless environment.

10.3 TCP Operations

When people refer to the 'bursty' nature of the Internet, they are not implying that everyone is choosing to access a web page or a favourite CD track all at the same time sending an avalanche of requests. In the same way as might be experienced by voice networks when a freephone number is given out to a radio or television audience, there will be occasions, such as pop concert broadcasts on the Internet, when an unusual surge will take place. For the most part, the bursty nature is caused simply by TCP doing its job.

The currency (or protocol data unit of measure – PDU) of TCP is a *segment*, and congestion management is handled through its modes and window sizes. TCP was designed to guarantee through its use of timers, congestion indicators and retransmissions that traffic would eventually reach its destination in the right order and occupy as much of the bandwidth resources as it could be provided with.

In the realms of real-time applications, those same guarantees, which come at a high cost in overhead bytes in each TCP header, have relegated TCP to providing initial call setup before handing off to other protocols such as UDP and RTP which are more lightweight, less complex and are more suited to the task (in the case of RTP). Interactive traffic is considered as an example of TCP's struggle to juggle its heavyweight headers and its acknowledgement procedures with the need to provide bandwidth-efficient and timely delivery for real-time applications carried over low-bandwidth links.

10.3.1 TCP Modes

In an effort to ensure that maximum throughput is achieved when transferring data, TCP is designed, in *slow start mode*[4], to double its sending rate with every consecutive transfer until it receives three duplicate ACKs for a previous transfer, thereby indicating packet loss between sender and receiver. Why three? This is because it cannot trust a single duplicate ACK in the event where a packet has been re-routed and is late in arriving. TCP puts more faith in its sequence numbers, which, if received out of order, require the receiver to confirm its last good sequence number. If the errant packet finds its way through with some delay, the whole sequence is confirmed, and all is well. If the packet is lost, a third duplicate ACK is received, and TCP enters *congestion avoidance mode.*

TCP's response to this is to reduce the sending rate by half initially. As there is a good chance that ACKs will still be returning to the sender with their RTT information, it uses these to set a new window size and stores this as its new threshold of congestion avoidance, which it calls the congestion window or *cwnd.* The threshold value is termed *ssthresh*, and the whole response is called a *fast retransmit.* For a fuller review of this response, see Stevens [35].

In the event that no ACKs appear at all, a *retransmission timer* kicks in, and everything falls back into slow start mode with a single segment being transmitted. The difference this time is that the *ssthresh* value is remembered, and TCP probes more cautiously as the threshold approaches. As no timing information (in RTTs) is available from the ACKs, another algorithm [34] smoothes out values returned before and after the outage to preserve some semblance of normality in a network that experiences great variances, even under 'normal TCP working conditions'.

10.3.2 TCP Window Sizes

In the initial three-way handshake, a sender's TCP stack learns of the receiver's *maximum receive segment size* and compares this with the sender's buffer size and the receiver's window size, taking the minimum of all three as the sender's window size. A 'window' in this context defines a TCP host's capacity to retain unacknowledged data that have been sent and at the same time have room to send volumes of data that will not inundate the receiver. The window size is minimally the addition of the sender's buffer and the receiver's *advertised window.* The size of these windows and of a host's buffers is a critical limiting factor in performance over WANs and is often overlooked in the initial setting up of a host system.

[4] The starting value historically has been 1, owing to the capacity of the buffer in early NICs. RFC2414 suggests values of up to four times the maximum segment size of the receiver.

Optimally, the receiver's window size and sender's buffer should be set to a minimum of *delay-bandwidth product* of the network path, i.e. the bandwidth in bytes per second multiplied by the RTT in seconds.

10.3.3 TCP and Interactive Traffic

Truly, TCP was made for bulk data transfers. Special measures were specifically created to cope with interactive traffic, such as Telnet and Rlogin, which involve many small packets and consequently would cause an ACK to be issued after every character was sent (or *echoed*) to screens as part of a terminal session. *Piggybacking* ACKs with data packets was one such measure, and *delayed acknowledgement* another, where an ACK could be delayed for as much as 200 ms to await the accumulation of more data prior to despatch. The overhead in doing this is not insignificant, a minimum being 80 bytes of protocol overhead to every 1 byte of data sent to the server and its corresponding echo. On an Ethernet LAN, this can rise to 120 bytes for every 2 bytes sent! This would be an intolerable waste of bandwidth on WANs and now in wireless networks, where capacity is at its scarcest.

The Nagle algorithm was designed to offset the effects across WAN links, by prohibiting further small segments to be sent until previous small segments have been acknowledged. Effectively, after the very first character has passed, others are grouped together and a single ACK piggybacked on to the returning echo. This naturally creates a variable delay, which, for character-based terminal sessions, is of no great consequence, but for jitter-sensitive applications, this algorithm cannot be used and is turned off.

Inevitably, a more efficient protocol was sought to supply real-time applications without the baggage of the TCP overhead, and these were found in UDP and RTP, which we now summarize below.

10.4 UDP Operations

User datagram protocol (UDP) is essentially a 'no-frills' transport protocol. It makes no guarantees about safe or sequenced delivery of the data, and has no congestion awareness, rate control, or QoS associated with it. It has no manners when in the presence of TCP, continuing unabated and unflustered by traffic bursts often adding to the congestion situation, while TCP politely backs off into slow start mode, and yet, UDP has been chosen as the transport mechanism of choice by real-time application protocols.

It is exactly this simplicity that makes it right for the task. Latency-sensitive applications like voice and interactive video would rather have 85–90% of their data delivered on time than 100% arrive with variable delays. These types of applications are sometimes referred to as *inelastic* as they value timeliness over packet loss. *Elastic* applications, however, like all their data to arrive in order and without loss.

As there are few examples of multimedia applications tolerances in respect of packet loss, it is worth considering the comparison of voice tolerances over various media available today [21][5] presented in Table 10.1. It is generally considered that voice delays of more than 150 ms are unacceptable because the human ear interprets silences as a negative response, or callers are constantly interrupting one another.

[5] A bandwidth of 56 kbps referred to for voice is from the US standard. The European equivalent would be 64 kbps.

Table 10.1 Comparison of voice connection characteristics

Connection type	Packet loss (%)	Audio stream (kbps)	Latency (ms)	Jitter (percentage variance in delay)
Normal phone line, local call	0	56	10–30	0
Undersea transatlantic link	0	56	100–200	0
GSM digital wireless call	5–15	13	100–300	50
Internet telephone application[a]	5–20	20	120–240	150

[a] Assumes a good ISP and dialup modem connection.

This table serves to remind us of two things: it confirms first our acceptance and tolerance of wireless networks and of their imperfections today and, second, the level of respectability, with the exception of jitter, attained by packetized voice over the Internet.

What can be said is that what UDP lacks in adaptability to network conditions and service level guarantees, it makes up for in its simplicity. When bandwidth is in short supply, as it is in wireless networks, that is a sacrifice worth making, provided that the applications themselves, or the protocols on which they rely, take care of the rate control, sequencing of packets and, most importantly, can adapt to network congestion conditions to provide differentiated levels of service. It is interesting that the emergence of the real-time protocol (RTP) used by multimedia applications takes the blurring of boundaries between application and protocol stacks, begun by the likes of ICMP's ping and SNMP's set, one step further. It is intended to be an application-visible, end-to-end protocol, whose processing is seen to be part of the application.

10.5 RTP Operations

RTP is not a rigid functionally complete protocol, but one that can be adapted to serve different applications' needs. This is achieved by carrying application-specific information in the RTP payload, using the *payload type* field to identify the format of the payload data (i.e. which application is using it) and the way in which those data should be interpreted.

A scenario for distribution of audio and video streaming sessions is given in Figure 10.4. The *mixer* component is often the RTP streaming server, whereas the *translator* can be a router replicating multicast[6] to unicast, or an application-level filter in a firewall.

All packets in RTP have sequence numbers to detect lost, out-of-order, or duplicate packets, and a timestamp field allows the correct re-assembling of data, preserving the timing of the original data. The *synchronization source identifier* (SSRC) shows which device is used for synchronization, and the *contributing source identifier* (CSRC) lists the sources joined by

[6] Multicasting is a bandwidth-efficient way of copying the same packet to multiple destinations. Routers with multicast software keep a copy of the packet and invite hosts from attached LAN segments to join a multicast group by means of an Internet group management protocol (IGMP) query. This contains a class D multicast address (in the range 224.x.x.x–239.255.255.255) to which interested parties respond with a 'report' message.

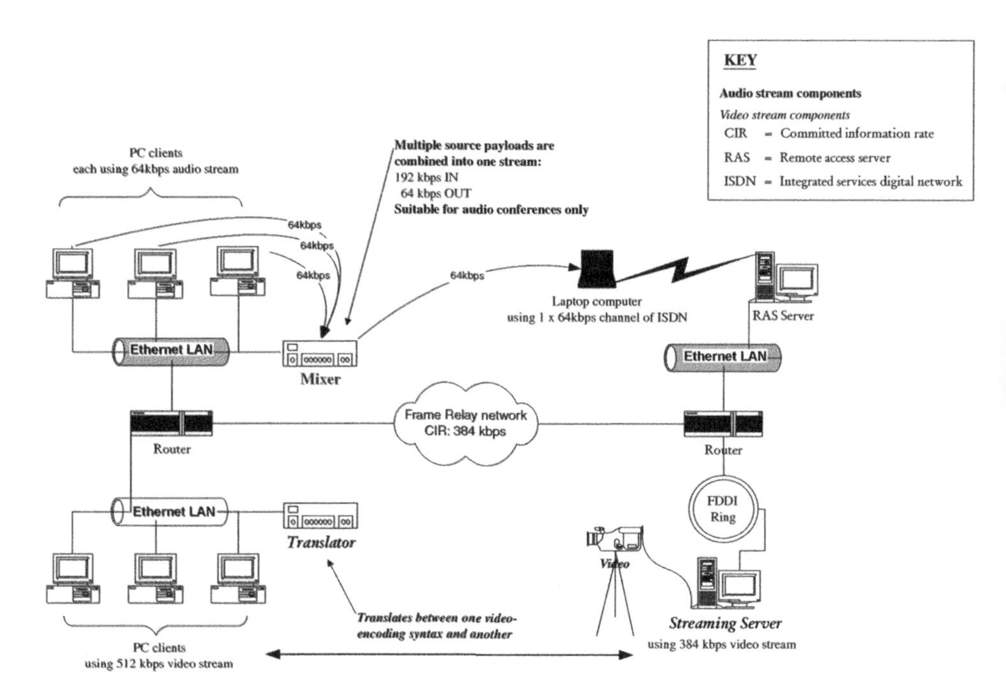

Figure 10.4 RTP streaming scenarios.

a mixer and notes the original SSRC. Once streams are mixed, the mixer becomes the synchronization source.

The SSRC is important not just for its rate control in playing back video streams, but because it can be used to identify individual application flows. This would be useful for applying QoS per flow. Regrettably, neither RTP nor its sister control protocol, the real time control protocol (RTCP), has any explicit QoS responsibilities assigned to them.

The RTCP allows information to pass between senders and receivers on the quality of the data in real-time. It also shares current session membership information among the participants.

In dispensing with the cumbersome TCP procedures, RTP freed itself to be able to interact with any application that requires synchronized and ordered delivery of packet streams. The cost of this is in the header increase, which, when added to the IP and UDP headers, amounts to 40 bytes. This is not as much as TCP, admittedly, but the problem of payload to header ratios still persists. Any increase in payload to compensate for the large header results as ever in increased jitter. Moreover, with no control mechanisms to provide QoS on the application data flows, we must look elsewhere for their provision.

With this in mind, we return to the question of how we provide a guaranteed level of service consistently when passing from fixed networks, on the one hand, armed with queuing and packet discard mechanisms, and not-so-slimline header to payload ratios, to jitter-prone wireless networks, on the other hand. One answer is to be found in a not-so-new standard, now undergoing something of a revival: IPv6.

10.6 IPv6

We saw that one of the mandated standards from the 3GPP recognized the need to introduce IPv6 into the IP multimedia subsystem (IMS) of the UMTS architecture. A conceptual model of how the UMTS Release 5 network will look based on current thinking is presented in Figure 10.5.

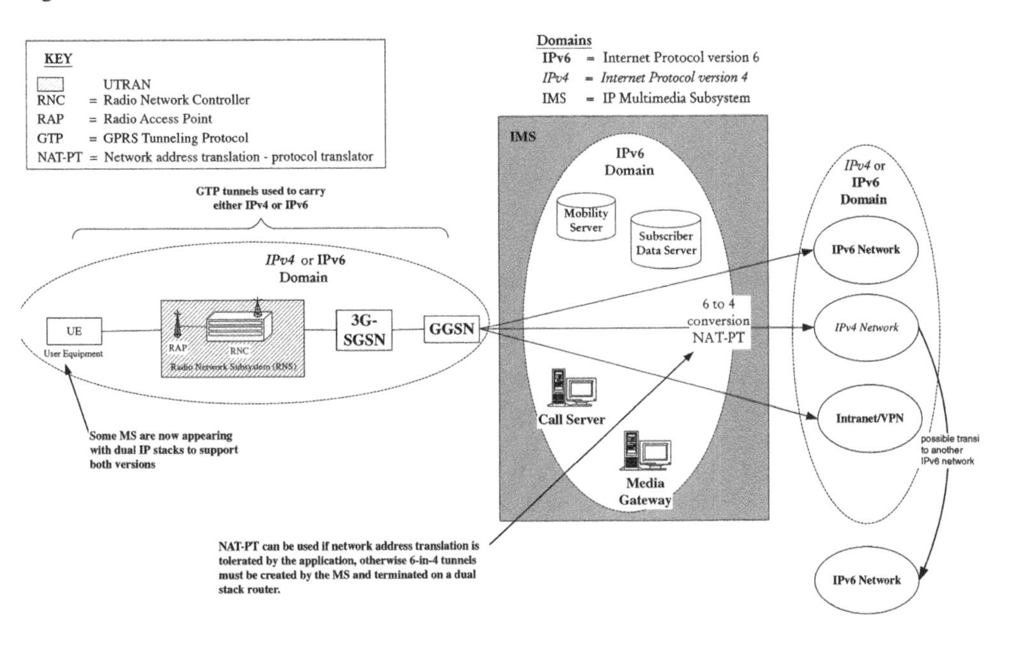

Figure 10.5 An all-IP conceptual network model for UMTS Release 5.

The IMS portion of this model is still under development but is expected to comprise call servers responsible for call setup to what will then be SIP[7]-enabled devices (an IP phone is an example of such a device), media gateways and servers responsible for subscriber profiles and mobility management.

This has galvanized vendors and manufacturers alike to dust off the cobwebs on this technology and give it a fresh airing.

Originally designed purely to provide more address space to an exhaustible supply of IPv4 addresses, it was also decided to tidy up the protocol with the intention to migrate from a data-specific protocol to a multiservice protocol [33]. The dearth of address space is felt most severely in those parts of the world, such as Asia, dependent on Class C Internet addresses. One global organization, for example, is said to hold more Class C addresses than the whole of China!

The introduction of IPv6 into the wireless infrastructure presents interesting possibilities, not because it contains explicit QoS support (which it does not) but because it throws off some

[7] Session Initiation Protocol is used to initiate packet-based telephony with minimal overhead.

of the inconsistencies and baggage that IPv4 had developed over its 20-year existence. The improvements are:

- the header has been reduced to a fixed size (40 bytes) and format (see Appendix B);
- per hop fragmentation is a thing of the past, owing to dynamic MTU discovery;
- options fields used to identify other encapsulated protocols are replaced by extension headers to keep the header to a fixed size;
- traffic class and flow label fields have been added;
- hierarchical address scheme to aid identification and speed up routeing:
  - Prefix (provider-based addresses)
  - Registry ID (allocating authority)
  - Provider ID
  - Subscriber ID
  - Subnetwork ID
  - Interface ID (host address on the Subnetwork)
- provision of a 'care-of address' for mobile users via address auto configuration.

This last point eliminates the triangle issue (see Section 9.2.3.3 Connection Management), embedding the home address within the care-of address for customer billing and inter-carrier cross-charging purposes.

The traffic class field began life as a priority field, but quite sensibly, it was recognized that technological advances would soon surpass any rigid definitions for traffic types, and there was also the need for compatibility with IPv4 packet classifications. At this point, it is likely that the DiffServ architecture (see Section 10.8 below) will be adopted as the supporting classification structure, while there is scope for the QoS control mechanisms to be incorporated into the flow label field.

IPv6 then is not a radical departure from IPv4, but a rationalized and simplified evolution of it with some additional functions specifically designed to support multiservice networks. We now look at the supporting protocols and architectures that will enable guarantees of service level performance to become more of a reality.

10.7 QoS Approaches in Fixed Networks

As we saw in Section 3.2 Router Scheduling and Queuing, the journey that IP packets takes can traverse many different network topologies with responsibility for their safe passage entrusted to either layer 2 (link layer) or layer 3 (network layer). It therefore follows that, in order to guarantee a consistent level of performance across all topologies, there must be either a single QoS response understood by all topologies at all layers from end-user equipment to back-end server or, more realistically, a transferable way of signalling that response end to end.

In this section, we will look briefly at the history of QoS and then look at two protocols gaining support in constructing a comprehensive QoS architecture, RSVP and DiffServ and a transport mechanism, MPLS, that may facilitate QoS in IP networks.

10.7.1 QoS Background

Historically, service quality fell into two distinct camps: those favouring class of service (COS), which defined a few discrete packet classifications interpreted at every hop along the route and those preferring the QoS approach, based on negotiated paths through a network with dedicated network resources assigned on a per-flow basis.

COS soon became IP types of service (TOS) after the field of the same name was used to identify TOS, with the first three bits indicating relative priority (or *precedence*) of a packet, and the next four (the *type of service*) stressing the importance to maximize throughput and reliability or minimize delay and cost. The eighth bit always had to be zero. These eight bits were hijacked by the group working on *differentiated services* (DiffServ) to provide three basic classes: *best effort*, *assured flow* for streaming application flows and *expedited flow* for high-priority and delay-sensitive flows, such as VoIP. Alas, in the conversion from IP TOS to what are now known as *DiffServ code points*(DSCP), the bit patterns are not interchangeable and therefore not fully compatible with each other (see Appendix C).

Out of the QoS branch came the *integrated services*(IntServ) group, which was accompanied by the Resource ReSerVation Protocol (RSVP), a protocol for negotiating network resources and by a transport method known as Multiprotocol label switching (MPLS), which, amongst other things, can carry embedded QoS parameters in the flow label field of the IPv6 header.

It is not intended to give a complete explanation of all priority and service class systems here. To convey the complexity and enormity of the task, Table 10.2 taken from Croll and Packman [21] illustrates adequately how disparate these architectures are. The baseline gauge used is the LAN priority system, as defined by the IEEE (802.1Q/p), which is then compared with alternative systems from the ATM Forum (ATM QoS) and the IETF (CoS, IPTOS, DiffServ and IntServ groups). Small wonder, then, with so many standards bodies involved that end-to-end QoS has struggled to get off the ground.

As we saw with wireless standards, successive re-definitions of traffic classes have to take account of current technologies, and new innovative ways of presenting information and fixed networks have not been immune from this phenomenon. The change from IP TOS to DS code points requires re-marking if the type of service field is to be translated (Appendix C). Interestingly, a second proposal for the use of the DSCP defined five bits for traffic prioritization and assigned a discard priority, similar to the DE bit of Frame Relay, to the sixth bit. The IPv6 specification does aim to mirror the DSCP reference model, but unfortunately, the discard bit proposal did not survive.

10.7.2 RSVP and IntServ

Making sure that there are enough network resources to fulfil any QoS contract is essential. The way in which this is accomplished is not yet agreed upon. At what point are the packets marked (application, host or desktop or now mobile, network edge), which part monitors the performance and downgrades or negotiates a downgrade in service under congestion conditions, and which element decides when to discard packets when saturation occurs? In searching for a comprehensive QoS architecture, the two camps described above need to be joined by a third: the desktop or host. So, the three approaches are as follows:

Table 10.2 Comparison of priority and service class systems

802.1p Value	802.1p[a] category	802.1p IntServ mapping	ATM traffic class	IPTOS precedence value	RSVP or IntServ flow type	IPv6 value (loose match)	Also known as	Application examples	Traffic profile
7	Network control	Network control	CBR	Network or Internet control	Guaranteed	Internet control (routeing, SNMP)	Assured service	RIP, OSPF, BGP4, SNMP	Critical to network health
6	Voice	Delay sensitive <10ms	RT-VBR	CRITIC/ECP	Guaranteed	Interactive traffic (Telnet, rlogin)	Premium service	NetMeeting, audio, VoIP, CBR video	Latency and jitter sensitive; low bandwidth; no self regulation
5	Video	Delay sensitive <100 ms	Non RT-VBR	Flash override	Controlled load		Rate controlled	Picture-Tel, Indeo	Jitter sensitive; high bandwidth; self-regulating
4	Controlled load	Delay sensitive	ABR	Flash	Controlled load	Attended bulk transfer (FTP)	Rate controlled	SNA transactions	Latency sensitive; predictable response times
3	Excellent effort	Reserved	UBR	Immediate or priority	Best effort		Assured service	SAP, SQL, critical business applications	Tolerate delays
2/0	Best effort	Default	UBR	Routeing	Best effort	Unattended data or not characterized	Best effort	Web browser traffic	Tolerate latency
1	Background	Less than best effort	UBR	Routeing	Best effort	Filler traffic (NNTP)	Default	Default	Default

[a] The IEEE redefined the 802.1D standard to include a 32-bit extension to the header to support virtual LANs (VLANs), referred to as 802.1Q extensions. Within 802.1Q, 3 bits (i.e. 8 values) are set aside for prioritization under the 802.1p extensions for traffic classification, but there is no facility for marking packet discard eligibility.

- The host or application marks each packet, using IP TOS or similar, with the desired service demands, and then the network is requested to honour them. This, I will call the *GQOS*[8] *approach*, as implemented by Microsoft[9].
- The application signals a request to the network for resources, which is accepted or denied according to current network load. This is the *IntServ approach.*
- The edge of the network decides which service packets receive according to pre-determined policies, aggregates packets with like requirements, and determines the amount of traffic admitted within each service class. This is the *DiffServ approach.*

RSVP is a supporting protocol for the IntServ architecture, which concentrates on the requirements of real-time applications and their transmission protocol, RTP[10]. IntServ assumes that resources must be controlled if there is to be any hope of delivering service quality, and to this end, some form of admission control should take place. Also, feedback information on the route's current status needs to be monitored and propagated to other network equipment. IntServ provides for two service levels:

- *Controlled-load* service – a 'better-than-best efforts' service for jitter-tolerant applications;
- *Guaranteed* service – a 'predictable' service, rather than the guaranteed service the name implies, that has upper limits on delay for jitter-intolerant applications.

With the inclusion of support for RSVP in local hosts, this makes RSVP a strong candidate for end-to-end resource negotiation and reservation. We saw how TCP (Section 10.3 TCP Operations) relies on a feedback loop to regulate send rates. RSVP goes one or two steps further by being able to embed current network resource status in its messages on a hop-per-hop basis, allowing for re-negotiation of QoS parameters between host and network, thereby enabling it to deliver the possibility of dynamic QoS.

This comes at a price, however. By definition, any protocol that is dynamic in nature must be able to change rapidly and is known as a 'soft state' protocol. To enable it to detect discrete and not-so-discrete changes in the route taken or be notified of router queue thresholds approaching exhaustion, regular updates carried in the *Adspec*[11] element of PATH messages need to be gathered from participating routers and switches (or *nodes*) along the path. The need for highly frequent sampling, especially where concentrations of high-bandwidth, high-speed links occur, ultimately will flood the network and therefore makes RSVP unsuitable for large global networks and particularly their cores. One suggestion was to piggyback the updates on OSPF 'hello' messages as they were doing their rounds of the routers. Perish the thought! The amount of traffic they already generate, especially in the presence of unreliable (or 'bouncing') links, is not inconsiderable.

---

[8] Generic QoS is a Microsoft definition, supporting both RSVP and DiffServ from Windows 98 onwards when the QoS Packet Scheduler is enabled (see Appendix C).

[9] Microsoft™ is a trademark of Microsoft Corporation.

[10] The Real time protocol (RTP) provides timing controls to other protocols such as SIP for IP telephony and H.323 for video-conferencing, ensuring that packets arrive in the right order and with the correct inter-packet delays for reliable playback.

[11] Adspec contents are divided into fragments, each associated with a specific control service. It is the ability to add multiple and new service classes in the future without altering the underlying transport mechanism that makes RSVP so attractive.

So, it would seem that RSVP is not *per se* an answer to the Internet's problems at the core of that network, but it may sit more comfortably at the network edge, where it could exert admission control, negotiate and adjust service levels with applications directly and provide feedback to the hosts.

RSVP was also designed to meet the needs of multicast applications, in that it is the receiver who dictates the QoS level supported at the end-user device. The receiver sends a RESV message in response to a PATH request from a sender. The former contains the flow information (flow type, QoS service and control required) or *flowspec* and can be altered by any intermediary node that finds that it cannot meet the resource requirements. In this way, large multicast groups can be serviced more efficiently, and dynamic membership of those groups becomes much easier to handle.

The PATH and RESV messages continue to flow after the application has begun, and rather like 'hoppa' buses, they stop off at routers on the way to glean any additional information they need. It is the flexibility to extend what type of information can be retrieved in the messages that has made it a popular, standards-track choice for negotiation and maintenance of service levels.

When deploying into real-world scenarios, it is also key to exercise an element of access and accounting control to define who is authorized to use these QoS services and identify who can be charged accordingly. This comes under the heading of *policy control*. While RSVP allows for the transport of policy messages as an Adspec object, it does not contain policies themselves. Policy control is a topic all on its own and so will not be covered here. For the standards bodies, this remains unfinished business within the IETF's RSVP Admission Policy or RAP group.

10.8 DiffServ

To address the need for a scalable solution for ISPs with large networks that did not consume router resources maintaining per flow reservation states, an alternative architecture was proposed, which borrowed heavily on the IP TOS per packet marking scheme while expanding the admission control mechanisms to include traffic classification and conditioning. These last two will be the focus for this section as they provide the means to combine application flows with differing levels of service without one starving the rest of the resources.

We remember that the IP TOS approach required packet marking at the host, which then passed unhindered through a 'dumb' network. DiffServ defines that this responsibility now lies firmly with the network at its edge or at the *ingress* to the network. The architecture comprises three main components:

- initial classification and conditioning of the traffic;
- writing of a DSCP into the DS field (or traffic class field in IPv6) of an IP header;
- per-hop behaviours (PHBs) triggered by the interpretation of the DS field.

By pushing the intelligence and decision-making to the edge of what now becomes a *DS domain*, an IP network could retain its 'pseudo-dumb' nature. All the core then had to do was interpret the DS field and execute a known, configured PHB to forward the packets.

The intention was to separate out the various processes and make them largely independent of one another. In so doing, an anomaly now arises, whereby one ISP's configured PHBs may not resemble another's in response to the same DS field. This effectively means that the quality of service applied at the ingress of the network may not be the same at its *egress* (or exit from the network). To continue the bus travel metaphor, while promising not to extend it further, the coat you took with you may not be the one you leave with, and what is more, it may not keep you dry.

10.8.1 Traffic Classification and Conditioning

Both classification and conditioning take place at the networks edge. *Classification* groups together like-flows into *aggregated flows*, which essentially share one common codepoint (DSCP), have similar content in their packet headers, and are travelling in the same direction. Like RSVP's requests for resources, DiffServ's aggregation is unidirectional. The *conditioning* element exercises some control over the amount and type of traffic permitted to cross a DS domain according to a set of rules known as the Traffic Conditioning Agreement (TCA), which are in effect SLAs set up between the service provider and its customer and in which an application will have a *traffic profile* associated with it. Part of the regulation of traffic flows through a DS domain, including drop priority, relies on determining whether the traffic arriving is in or out of profile.

Conditioning comprises four tasks as follows:

- Metering – measures traffic rates to determine in- and out-of-profile traffic;
- Marking – the setting or changing of code points (DSCPs);
- Shaping – controls traffic emission and ensures profiles are adhered to by aggregate flows;
- Policing – discarding traffic.

As a result of the metering, the decision to mark, shape or drop will be taken. The marking not only sets the DSCP but also adds the packet to a DS aggregate flow (or *behaviour aggregate*). The packet can also be marked to a set of code points, which are mapped to different per hop behaviours to deal with changing network conditions as perceived by the meter. It must be remembered, however, that the meters only know about traffic entering their edge of the network and not about the network as a whole.

Dropping packets can be seen as a form of shaping them when the flows do not conform to a traffic profile. The shaper will store a stream in its memory to enforce compliance to a TCA, but will drop packets if its memory is exhausted.

10.8.2 DiffServ Marking

Returning to the three approaches to an adequate QoS Architecture (Section 10.7.2 RSVP and IntServ), the question arises as to who exactly will mark the packets with its DSCP and, once marked, who will honour the DSCP and who will ignore or re-mark it.

End systems are quite able to mark packets under the GQoS approach, but network and service providers do not trust end devices or their users. However, the packets could be policed at the network ingress point, checked for validity against a known set of TCAs and forwarded or downgraded as necessary.

A second way would be to assume nothing about end devices, ignore any DSCPs, and re-mark according to network admission policies and traffic classification profiles.

A third way would see end systems using RSVP to set up the flows across a DS domain, trying to approximate the RSVP traffic classes to DSCPs or RSVP admission policies deciding not to use a DiffServ marking at all.

As DS code points were discussed in Section 10.7.1 QoS Background and in Appendix C, we will move on to see how those code points (DSCPs) are interpreted by the routers or switches in the network.

10.8.3 Per-hop Behaviours

As we saw in Section 10.2 Router Scheduling and Queuing, the settings contained in the code points must then be translated into mechanisms or combination of mechanisms that suit a particular application or its transport protocol. A per-hop behaviour (PHB) is the way in which router resources are allocated to aggregated flows. These can be referred to in terms of simple priority queuing or of link characteristics such as latency bounds, jitter tolerance, and discard precedence.

The 6-bit length of the DSCP field (see Appendix C) provides for a total of 64 possible treatments of packets or PHBs, to be determined by the network providers. It is important to realize that no core set of PHBs has been defined, assuming that a common set will emerge as 'supported' PHBs from an industry forum. Given the complexity of router queuing possibilities, the IETF's optimism may well be justified.

Similar to the IntServ traffic classes (which took the best effort class as read), DiffServ defines three PHB groups:

- *Class selector PHB* – a best-effort service aimed at providing some compatibility with IP TOS precedence fields;
- *Assured Forwarding (AF) PHB*– equivalent to a Frame Relay service that preserves the packet order, allows out-of-profile traffic to be carried with the same service as other traffic in the same flow or *micro flow*and drops flows that exceed their TCA according to a *drop precedence level*;
- *Expedited forwarding (EF) PHB* – equivalent of the 'guaranteed service' of IntServ to provide voice and video traffic with the low latency, low jitter and low loss characteristics that they require.

The first of these allows for any of the queuing techniques [priority queuing (PQ), CBQ, WFQ, or WRR] to be used to forward on a per-packet basis according to priorities set in the DSCP. This means inevitably that packets will be re-ordered according to the prevailing network conditions, which applications like TCP do not cope well with. An out-of-order packet can send all the wrong signals to the sending TCP application, committing it to a slow start mode prematurely and giving rise to more 'bursty' traffic.

Assured forwarding (AF) was designed specifically for such applications. It may drop packets within a micro flow but maintains the order sequence. AF contains 12 PHBs altogether, comprising four service classes and three drop precedence levels, with the allocation of buffer space and bandwidth being the distinguishing factors between the classes, as shown in Table 10.3.

The colour scheme is not my own, but that of a proposed admission marking scheme known as two-rate, three-colour marker (trTCM), which not only helps to understand the reverse logic of drop precedence levels but also serves as a useful pointer as to which packets would be dropped first (red), which might be forwarded perhaps under best-efforts (yellow), and which would go largely unhindered (green).

To satisfy the bandwidth and buffer allocations, one can see that either CBQ with defined

Table 10.3 Assured forwarding (AF) classes and drop precedence levels

Drop precedence	Class 1 (DSCP)	Class 2 (DSCP)	Class 3 (DSCP)	Class 4 (DSCP)
Low	001 010	010 010	011 010	100 010
Medium	001 100	010 100	011 100	100 100
High	001 110	010 110	011 110	100 110

levels for each service class or WFQ could act as the queuing mechanism, while weighted random early detection (WRED) could be used for the drop precedence levels.

The last PHB group, *expedited forwarding*, assumes little or no queuing delay with no interference from other application flows end to end. This presents interesting challenges for ISPs who *must ensure that incoming traffic into their DS domain does not exceed their network equipment*'s *capability to forward it on to all outgoing interfaces.* Furthermore, this applies not only to all of a single ISP's interior network nodes but also to all of the network nodes in adjacent ISPs who have negotiated reciprocal and chargeable service contracts between them.

Priority queuing could satisfy the EF requirement but must guard against denial of service attacks[12], which would take advantage of such unqualified prioritization. CBQ could be used as an alternative to limit bandwidth allocation but runs the risk of introducing increased jitter into the EF class.

Adherence to a strict admission policy is a prerequisite for EF classes to succeed, and this must be applied to the network edge or ingress point. This effectively shifts what I will call the 'jitter point' as well as the 'drop point' out to another part of the network. Video and voice services must be guaranteed end to end, which must be an inclusive arrangement with client systems, servers, and all points in between.

To address this need for end-to-end service quality and combine it with the need for extensible scalable routeing, a protocol was developed to accompany the aforementioned QoS architectures and facilitate a common transport mechanism that could cross the divide between fixed and wireless networks: MultiProtocol Label Switching.

10.9 MPLS

This section does not consider all the aspects of MPLS, which is in itself a broad subject, but seeks to identify those components that make the QoS-supported path from a fixed IPv4 network to an eventual IPv6 wireless network more of a reality.

We look at each of the following to determine its usefulness in a multiservice network:

- perceived benefits
- its fit with IPv6
- multiprotocol nature
- MPLS operations
- flexibility it holds for future solutions and development

[12] Denial of Service (DoS) attacks feed off a network's capacity to forward traffic at high speeds and a server's incapacity to service a garbled or corrupt request.

We will see how a middle way has been found to accommodate both camps (see Section 10.7.1 QoS Background), be they QoS or COS, IntServ and DiffServ, connection-orientated and connectionless or IP packet-based and ATM cell-based networks, combining the best of both worlds and making them better.

The duress that IP packet-based routeing was under to deliver QoS-guaranteed routes just on a single path in one direction has been relieved to a great extent by MPLS. One of the additional benefits this brings from the circuit-switched world is to be able to assign specific and alternative routes to applications requiring near-constant quality, queuing, and bandwidth resource in a dynamically changing network without the need to wait for route convergence[13], which is disruptive and a constant source of irritation in large data networks where non-aggregated[14], routeing tables can be considerable in size.

10.9.1 Benefits

The idea of MPLS is a very simple concept and one that has been applied in the definition of DiffServ, for example, in the way it promotes independent operation between components:

> ... it (MPLS) decouples the information used to forward a packet from the information carried in the IP header. It does this by using labels ... to make forwarding decisions. This simple fact has far-reaching consequences. [36]

The IP routeing scheme relies on looking up an entry in a routeing table to find the desired network and the interface that the packet should be forwarded on. This was fine in an IPv4 world where addresses were neatly defined into classes A, B, and C of known length. A search on an address starts with the leftmost bit and continues the length of the address until it finds the nearest or *longest match.* When address space was seen to be in short supply, this was overcome with CIDR[15], which made the look-up more complex and therefore more processor-intensive. Any change in the forwarding algorithm to cope with these changes is costly and time-consuming.

The prime benefit of MPLS for manufacturers is that the forwarding algorithm is fixed, and thus any new variant requires no change to the forwarding component. Other benefits include:

- IPv6-ready – no changes to the forwarding component necessary;
- support for QoS;
- fixed length labels – forwarding made easy;
- fast re-routeing around network failures – no waiting for re-convergence;
- simplified integration of ATM into IP networks;

[13] Convergence is the process of propagating a route state change out to all network nodes, which rely on a global addressing scheme for their routeing. In large networks, this can take from 20 s to several minutes to complete, during which time application activity is suspended, queues fill up, and packets are dropped.

[14] Aggregation or Route Summarization is a means of shortening the look-up time of a route through careful planning of addresses by location, country, and continent. This benefits only greenfield sites or those with well-organized IP address schemes.

[15] In OSPF terminology, this is referred to as variable length subnet masking (VLSM), which aptly describes how the lengths of host and network portions of class A, B, and, C addresses are no longer fixed. On serial point-to-point links, this reduces the normal 256 consumed host addresses down to 4, allowing the other 252 to be redeployed often on other serial links, making recognition and address administration much easier.

- traffic engineering – allows for load balancing on links, switching dynamically to less congested routes;
- scalable virtual private networks (VPNs).

The most important benefits from a service level performance perspective may appear to be just the first two, but the others (which are not discussed here at great length) facilitate consistency of performance in congested periods and scalability of the solution.

Looking up labels is a much less complex task than that of IP routeing and is therefore much faster. The labels also only have local significance, i.e. they do not rely on holding a vast amount of information about an entire network, but use labels instead to identify an incoming packet and the interface or port on which it is to be forwarded. The labels are unique only to the network node and can be reused at the next hop in an entirely unrelated manner. What we must stress at this stage is that MPLS is *not* a replacement routeing protocol, but a transport protocol. It is dependent on layer 3 information for its routeing, but has the added advantage of having alternative routes already set up in a link layer2-switching way, which it can switch to at a moment's notice. It is this approach that provides the fast re-routeing and traffic engineering qualities of MPLS.

VPN customers can look forward to label identifiers specific to their own routes, so that they are not routed to another VPN customer and support for overlapping or non-unique IP address usage, avoiding conflicts in address clashes between customers.

10.9.2 Fit with IPv6

The IPv6 flow label field (see Appendix D) is of particular interest, because it allows for applications flows to be identified for special treatment in a QoS way, when accompanied by an IPv6-enabled reservation protocol such as RSVPv6. This means effectively that flows with known QoS requirements can be identified by a flow label and may even be aggregated with other flows with matching labels in a DiffServ-favoured aggregation manner. Although as yet undecided, the Traffic Class identifying different types of traffic seems to favour mirroring the DiffServ classifications. Given the propensity of traffic classes that we have seen already, this may be an eminently sensible view to take.

10.9.3 Multiprotocol Nature

The MPLS label sits like a *shim*[16] in front of the IP header and before any subsequent encapsulating header or link layer header in a layer sometimes referred to as Layer 2.5. It is therefore in the useful position of being agnostic of, and independent from, both layer 3 protocols (IP, IPX, AppleTalk) and layer 2 link layer protocols (ATM, Frame Relay, Ethernet, PPP). This independence is its strength and has earned it the label of being truly multiprotocol in nature.

10.9.4 MPLS Operations

The framework for MPLS mandates that it must interoperate with RSVP. This is essential, as

[16] In the case of ATM, it supplants the virtual path identifier (VPI) and virtual circuit identifier (VCI) fields.

MPLS is not an end-to-end service guarantor, because labels are only inserted at the network edge by a label switching router or LSR. This does not mean that it cannot interoperate with DiffServ and, indeed, one proprietary approach known as *tag switching* simply substitutes the IP TOS precedence fields directly into the Class of Service (CoS) field of the MPLS label (see Appendix E).

When RSVP is used to reserve resources for a flow and its associated traffic class, a label switched path (LSP) is set up across the MPLS domain, and the traffic passes along it, changing or *swapping labels* from inbound interface label number to outbound label number at each hop. In order that it may be routed correctly, there is a label-to-network prefix association in the forwarding tables for multiple routes to a destination network. If a network failure occurs, the change of state is reflected in a change of label-to-network prefix to a secondary route. The key point is that it is the LSR that chooses the path through the network based on the information provided to it by routeing protocols or potentially from any other sources.

Those of us with at least one eye on application sensitivities and response times realize that waiting for a break to occur is not the optimum recovery method. Far better to be warned of the impending congestion and have a threshold breach signal a change of label request, than to wait for the inevitable.

Once a labelled flow reaches its last LSR at the egress point of the MPLS domain, the label is removed, and the normal layer 3 routing delivers it to its destination.

10.9.5 Flexibility and Future Development

In order to cope with multiple MPLS domains, it is possible to stack labels so that only the outermost label is read for example by the border gateway protocol (BGP4), which is a routeing protocol commonly used by ISPs between OSPF autonomous systems (AS). Extensions to BGP4 that have been developed are there to accommodate MPLS labels.

Another proposal[17] changes the original remit of RSVP from one of resource reservation from destination host along a defined routed path to one of requesting forwarding as well as state information on collections of flows outside of the chosen route. Effectively, pairs of label switched routers (LSRs) as well as end-host systems can negotiate paths (LSPs) with different QoS characteristics.

It is because MPLS has no strong affiliations with any particular architecture or protocol that has gone before that we have the freedom to develop different applications for it. As we have seen with ATM, MPLS can sit as comfortably on a layer 2 switch as on a layer 3 router or, in other words, with both acting as LSRs. This opens up enormous possibilities for future development since a device does not have to automatically recognize IP packets to qualify as an LSR in an MPLS network. Labels could be represented, for example, by wavelengths or *lambdas*in optical networks, and this standardization is underway under the heading of generalized MPLS (GMPLS).

One can conclude from this that MPLS is an important piece of the jigsaw for facilitating the transport of QoS requirements across multiple domains and the re-routeing of the packet flows in a non-disruptive manner. When accompanied by the 'helper' protocols

[17] Provider Architecture for differentiated Services and Traffic Engineering (PASTE) is part of RFC 2430, proposing RSVP at the edges of networks with MPLS at the core.

and architectures for negotiating and setting up QoS levels from end-user devices, backed up by the appropriate queuing mechanisms along the route, one can see that extending their use into a wireless network is not without the realms of possibility.

Although not all has been settled in the standardization process of all components considered above, it is certainly true that MPLS has provided the delivery mechanism for QoS in multiservice networks. The issues that remain revolve mainly around the maintenance of that performance level once established. To achieve dynamic re-routeing, for example, of which MPLS is quite capable, one must have the correct information fed to it from perhaps more than one source reliably, the source itself having to be governed by the same QoS guarantees as the flows themselves.

One could easily argue that architectures and protocols alone would not have been enough without the disarming practicality of being able to condense whatever data or voice traffic one wishes to send into small enough packages that a wireless network can pass, error-check, and re-transmit, if necessary, without disrupting the application or adversely affecting the data, voice or video streams. This is the task of compression techniques.

10.10. Compression Standards for Multimedia Applications

In this section, we consider the contribution made by compression techniques and focus on those related to video applications such as NetMeeting (see Section 10.7.1 QoS Background, Table 10.2), which will be used in the lab tests discussed later. This remains a central requirement underpinning any transmission of multimedia applications but becomes more acute when traversing to a wireless network where bandwidth is at a premium.

We will look specifically at the different payload types that can be carried by RTP, examine the video standards in circulation, discover what to expect from header compression, and understand how video images are compressed.

It will be seen that, compared with the regular sampling undergone by voice traffic pumped out at a constant bit rate, compression introduces inevitable delays at both ends of a link where it is applied. The control over these delays will be more important to interactive or *conversational* video than to streaming or *playback*video in so far as buffering at the receiver can do much to relieve inconsistencies in the playback mode, but be less effective in conversational mode.

10.10.1 RTP Payload Types

Network equipment will look into the RTP Payload Type field of the RTP header to classify traffic and handle it accordingly. The various encoding techniques are given here for completeness, but the essential information for our purposes is the data rate or bandwidth expectations of each. The payload types are listed in Table 10.4.

The reasons for the fluctuations in MPEG standards is partly due to the need to address audio-only markets, as opposed to audio and video markets separately, as is the case with MP3 players now available to replace the cassette tape 'Walkman'. The upper limit reduction from 18 to 6 Mbps can be attributed to improved techniques for transmitting broadcast-quality television.

Exceptionally low data rates of 5.3 kbps for voice can be a mixed blessing. While admittedly not consuming as much bandwidth, 'noisy' lines experiencing even small data losses will result in larger portions of the voice packets being absent altogether than would be otherwise lost in a 64-kbps call, which may make conversations at best hard to follow and, at worst, unintelligible.

The unavoidable fact with compression is that *delay is inevitable*. A sufficient amount of data needs to have been gathered before compression algorithms can operate, and this introduces delay from the outset. For example, a voice sample[18] carried over ATM at 64 kbps would need to be split up into 48-byte payloads. After the first byte arrives, the other 47 bytes would take 47/8000 or 5.8 ms to complete before the header of 5 bytes is added and the ATM cell despatched. This delay is often referred to as *packetization delay* or *packet quantization jitter*.

Table 10.4 RTP payload types and data rates

RTP payload type	Data rate	Description
G.711	64 kbps	Standard voice encoding using PCM[a] (A-law and M-law variants)
G.723.1a	6.3 kbps	ACELP encoded dual rate voice
G.723.1m	5.3 kbps	MP-MLQ encoded dual rate voice
G.729	8 kbps	CS-ACELP encoded dual rate voice
GSM	9.6 kbps	Global System for Mobile communications (Europe and Asia)
H.261	128–384 kbps	Video codec[b] designed for video-conferencing optimized over 6 ISDN channels
H.263	20–384 kbps	Video codec for designed for desktop and wireless media conferencing
JPEG		Standard colour image encoding and compression
MPEG-1	100 kbps–1.5 Mbps	Standard for audio encoding used in digital storage media such as CD-ROM
MPEG-2	3–18 Mbps	Standard for audio and video encoding used for broadcast TV, digital video disc (DVD) and high-definition TV (HDTV)
MPEG-3	8–256 kbps	Standard for audio encoding
MPEG-4	20 kbps–6 Mbps	Standard for audio and video encoding used for multimedia applications, web authoring, 'wireless media phone'

[a] Pulse Code Modulation has two variants: A-Law is the European standard, and M-Law is the North American standard.

[b] Coder-decoder or codec refers to the converting of a video signal into packets (*packetization*) at the sender and back again at the receiver.

10.10.2 Header Compression

Headers can also be compressed. The TCP/IP header for instance has been replaced by a symbol using van Jacobsen header compression over slow serial links for some time using the point-to-point protocol (PPP). This can reduce the propagation delay by 10 ms over a 28.8-kbps modem. In an effort to reduce the packet quantization jitter, multilink PPP extensions took this one step further by interleaving fragments of larger packets with high-priority packets.

[18] At 64,000 bps and at 8 bits per sample, voice traffic is sampled 8000 times per second.

It depends very much on the link layer protocol as to the packetization delay encountered. A modem will experience between 20 and 40 ms of delay for packetization, whereas an ISDN terminal adapter will only experience 4–5 ms of delay. An Ethernet-to-basic rate interface ISDN connection only experiences 1 ms of delay at most [21]. This means that for some latency-sensitive applications that travel between two modems (such as Internet gaming), better performance would be achieved by turning data compression off but leaving header compression on.

RTP header compression, which includes RTP, UDP, and IP headers, can reduce the size of the header from 40 to just 2 bytes. This is naturally a vast improvement, but the real art is in configuring maximum MTU sizes so that the header-to-payload ratio is at its most efficient. A large payload (1500 bytes, say) and small header might seem attractive, but the protracted wait for all 1500 bytes to arrive would not make the wait worthwhile to jitter-sensitive applications.

Video application flows or *streams* have their own characteristics and compression methods, which we must understand to gain an appreciation of how their performance can be influenced.

10.10.3 Video Compression

To understand video and video–audio combination compression, it is useful to consider the different compression techniques available today. They are:

- Run-length encoding – good for regular shapes and monotone images;
- Algorithmic encoding – mathematically based suitable for colour images;
- Differential encoding – for compressing moving images;
- Audio compression – also mathematical, but uses silence suppression and multiplexing to enhance voice compression.

Run-length encoding takes repetitive bands of a colour, assigns a symbol to each band, which can then be encoded again to reduce the file size (see Figure 10.6). Such a technique is used in the formatting of GIF files using Lempel-Ziv Welch compression.

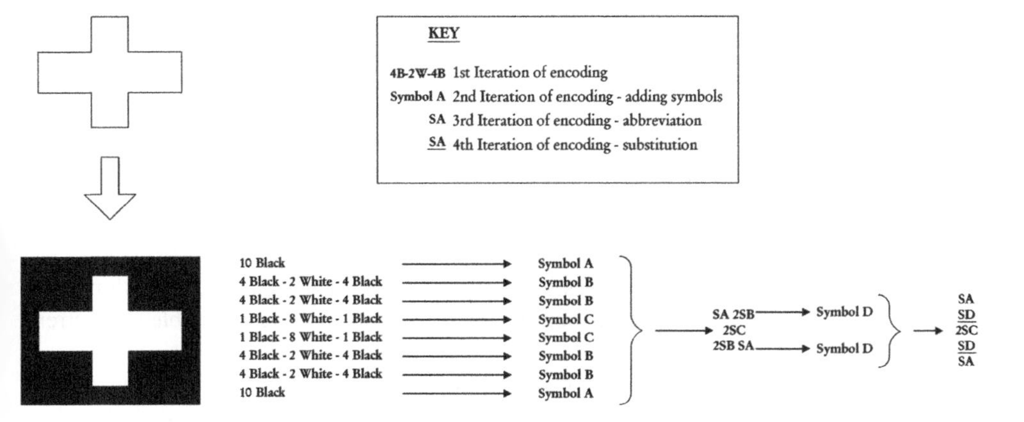

Figure 10.6 Run-length encoding.

Colour photographic image quality requires more information because of the nuances of colour shading (hue), depth of colour (saturation) and brightness (contrast). The JPEG format relies on gathering information on colour variance around the centre of an area and processing many samples in the area using *algorithmic encoding*. The fact that there is so much information to gather plays right into compression's hands. The more a compression technique knows about an image, the more it can assimilate that information and reduce its size.

A 2D image strangely has three dimensions: horizontal and vertical co-ordinates and colour depth. A 2D moving image is governed by the fourth dimension of time. Once again, there is sufficient information about an image, there is only a need to transmit the changes or differences rather than refresh the whole image. This is the substance behind *differential encoding*, which allows a normal broadcast television picture to be compressed from a hungry 6 Mbps down to several 64-kbps channels [21].

Differential coding creates its own bursts of traffic, by sending entire *key frames* that describe the whole image, as opposed to the *intermediate frames*, which send the changes or *deltas*. Under congestion conditions, any late arrival will be discarded by either the router or the receiver, resulting in the *clipping* of these key frames. The visible effects of this are incomplete or poor-quality image and synchronization between the receiver's ability to buffer video streams, and the sender's data rate may be lost. One solution to this is to make the key frames smaller by reducing the display size (resolution) or the number of colours. This makes the playback smoother, but any significant changes to the image will be slower to propagate, resulting in the fairly common scenario, over low bandwidth links, of the picture appearing to freeze while the audio commentary continues.

Finally, voice compression techniques have been well established, which take advantage of the large parts of a conversation that is silent in one direction or the other. By multiplexing other voice calls into the same packet stream, considerable efficiency gains are to be had reducing the normal 64 kbps down to 8 kbps and lower. The only two caveats with employing such highly compressed, low-bandwidth audio channels is first that any data loss means losing words or even short sentences rather than syllables, and second, that the distinguishing characteristics that make our voices individually recognizable begin to diminish below 8 kbps until they appear robotic in sound.

10.11. Summary

As we have just seen, compression is a necessity in most cases for transmitting over low-speed links. However, compression is only one of a package of measures that has to be employed to deliver continuous service quality for a single real-time application flow from end to end in both directions. Negotiating and selecting resources at video call setup is a complex operation and will include:

- application or end-user device requesting a certain service and packet classification level, which it expects the network to honour or adapt according to routes available and current network traffic levels;
- a TCP control channel initially with RSVP to find a route back from a destination, a route whose routers and switches have the necessary queuing mechanisms in place to translate traffic classes and IP TOS precedence fields into reliable service tasks, to include:

  - classifying or re-classifying traffic
  - inserting traffic into different queuing mechanisms or combinations of mechanisms
  - reading and setting packet discard eligibility
  - discarding time-sensitive traffic that is so late as to be useless
  - invoking selective discard of non-prioritized traffic when congestion thresholds are breached
  - dynamic re-assigning of buffer space to support flow control, adequate not to starve lower-priority traffic or allow longer flows to dominate queues
  - send appropriate feedback to applications that operate flow or rate control in times of congestion
  - forwarding packets along a label path via the appropriate interface

- negotiation of available service levels across DS domains between service providers or carriers with agreed billing levels verified;
- embedding of available QoS service levels into MPLS labels at the network edge;
- setting up alternative label switched paths to destination in case of network fire to enable dynamic switching of application flows with negligible interruption to service;
- confirmation of service level with appropriate traffic conditioning now in place to allow data traffic to proceed.

Once set up, the flows then pass to two UDP channels for the transmission of voice and data, timed and regulated by RTP. It is then assumed, though not yet defined, that monitoring of the service level continues through the services of RSVP in combination with other sources of information on current network conditions to feed MPLS with the dynamic information that it craves. It is also assumed that any QoS-related activity would recognize the connection between the TCP control channel and its dependent UDP channels and assign a QoS priority to an RTP-related TCP transaction over and above a normal TCP transaction.

In moving to wireless, most of these architectures need little adaptation to reach the 'radio network edge' at the SGSN, whether across an IPv6 domain or not. The multiple addresses, for instance, that mobile devices will be supplied with by neighbouring cells on a select-best-signal basis will need to be accommodated perhaps through the use of restricted multicast methods or through multiple LSPs to enable service to be requested by, and delivered to, roving IP devices.

We have very much concentrated on the needs of individually classified applications neatly boxed into service categories, but the reality of any network is that multiple flows must co-exist or be controlled in a fair and equitable manner for these well-intentioned QoS levels to be maintained. The next chapter focuses on performance tests that I have carried out to test a mixture of applications that a wireless service might want to provide as a bare minimum over the same air interface, and the conclusions that can be drawn from them.

11

Wireless Performance

In a world where UMTS or 3G networks are only just being built out and where we have yet to see the applications and services designed for that architecture, I felt it important to be able to quantify the response times that might be experienced in such an environment and therefore set the service level expectations of such applications and services when transmitted across a wireless network in order to gauge whether mobile IP devices can truly be a portal to multi-service networks.

In this chapter, we will first consider two other examples of testing that have been carried out using different models and methods and the conclusions they drew. After an overview of the simple test lab topology, we focus on specific applications and what happens to them under the 'stress conditions' of having other applications compete for the same bandwidth in a simulated wireless environment. We will look at each of the following:

- a GPRS simulator and its results;
- a 3G demonstrator and its results;
- the test laboratory, its methodology, topology and results.

The results included here are a representative sample of a number of tests carried out between October 2000 and May 2001 at two different locations. It is intended that the sample provided enough initial evidence to investigate some aspects more thoroughly.

We will see what efficiency gains were available in a data network that are unavailable in a wireless network, how jitter in a VoIP application is not tied to throughput on a live network, how videophone feasibility may be possible only at bandwidths over and above 128 kbps, and how colour could affect bandwidth requirements.

For the purposes of this testing, bandwidth restrictions of around 20 kbps were deliberately chosen as the best-case scenario under GPRS and restrictions of 384 kbps chosen optimistically as the best case under UMTS. Not surprisingly, we will see how transfers of 1 Mb, such as those carried out by FTP or e-mail attachments will be feasible but impractical for all but the most patient in a GPRS network. This may well serve to warn those wishing to download MPEG4 audio files to their PDA that they may have to wait.

We will also see how, under the stress of a 'bandwidth hog', a near real-time streaming or playback application (see Section 10.10 Compression Standards for multimedia applications) performs reasonably well in a simulated 3G network, when given enough bandwidth (384 kbps) to fend off rival applications. Delays are comparable with running a similar video and audio streaming application over a 56-kbps dial-up modem, which delivers no more than 35 kbps at the best of times.

11.1 GPRS Simulator

This in essence was a protocol simulator with adjustable parameters designed to investigate protocol interactions and to be able to test new standards and protocols as they appeared. It also looked at throughput and capacity and inferred achievable bit rates and delays.

The disadvantages of this approach are as follows:

- it requires traffic models for user behaviour;
- it requires traffic models for application characteristics;
- it requires models for aggregated numbers of users[1];
- the use of a physical layer model for the air interface in intensive protocol simulations is not possible, because of the computational effort required;
- statistical parameters were drawn from a dial-up modem.

The simulator [28] only looked at Internet web browsing as an application, as it was assumed that Internet browsing would be the primary application for GPRS data use. It assumed one timeslot on the uplink and looked at one, two, and four timeslots for the downlink reflecting the asymmetric nature of web traffic and refers to this multislot capability as MSC1, MSC2, and MSC4. The report made the following important conclusions:

- The CS-2 scheme is preferred under normal radio conditions and that CS-4 should be used in areas of high radio quality where forward error correction is not needed.
- As the number of web users per cell increases from 10 to 30, the multislot capability advantage disappears. Both MSC4 terminals (initial throughput of 22 kbps) and MSC1 terminals (initial data rate of 8 kbps) were reduced to a mere 6 kbps, above which limit the report describes to be 'an acceptable speed'.
- Efficient use is made by GPRS of the scarce radio resources, by using the idle time of some users with the data of others and multiplexing them together. A system capacity determined to be 66 kbps for CS-2 and MSC4 returned 300-kbps throughput for 20 users operating at a mean bit rate of 15 kbps.
- Users will experience a wide range of data rates.

It is clear from this and from the published data rates listed above (see Section 9.3.5 Coding Schemes, Table 9.2) that my selection of 20 kbps was a fair estimate as the best-case scenario for GPRS in the tests I undertook. However, the primary application of web browsing has been subsumed into the need to access most applications from anywhere anytime.

11.2 3G Demonstrator

Similar to the GPRS simulator, this [38] was implemented on a single system with a module to simulate the UMTS protocol stack this time and additional modules to represent an IP proxy server to enable real applications to be used to connect with the simulator and the physical layer (PHY-link) to inject errors based on scenario-driven radio-link simulation.

It describes a system architecture for applying one of four QoS traffic classes (see Section 9.4.2.1 QoS Traffic Classes), bandwidth and reliability requirements that revolve around

[1] An Internet usage model, for instance, is the nirvana of many laboratory technicians, which is often a fruitless and futile search.

retransmission techniques such as SR-ARQ (see Section 9.3.2 GPRS Lower Layer Protocols), and a future delay requirement to test scheduling mechanisms. At present, their scheduler works on a round-robin basis only.

While it states a goal of investigating 'user-perceived QoS' and reports that various concurrent TCP and UDP applications such as file transfer, web browsing, and a proprietary H.263 + + video codec have been demonstrated, regrettably, no results of those demonstrations have been published. It looks to broaden its field of investigation to RTP-based video applications in the future.

This is clearly a system where the 'driver' must enter different combinations of parameters and even the QoS parameters themselves to see its effects on a given application flow, whose port numbers are known. Unfortunately, this is a luxury with applications using RTP/UDP, as they use different random port numbers for their control and data flows, as we shall see.

11.3 Test Laboratory

Much of what has gone before, including other test work and the preceding chapters, has concentrated on single applications and their response or lack of it to congestion. Even the routeing and queuing mechanisms and the DiffServ and IntServ architectures consider individual flows or the finer granularity of per packet marking. The time has come to stand back from the detail and consider the more basic question: will today's multimedia and mainstream applications co-exist and perform over an air interface to a degree that is acceptable to most of their potential customers?

In this section, we look at the methodology followed and the simple layout of the test laboratory, before examining the test results. I have prefaced the tests carried out in the laboratory with those run across a live network to give an appreciation of how the types of application we are trying to run perform on networks today with limited QoS capability.

11.3.1 Methodology and Test Lab Topology

The testing system used relies on agent-generated traffic based on real-time captures of the application concerned. The flows are then analysed, converted into scripts (with the ACKs removed so as to represent pure data flows between the client, *Endpoint 1* and its server, *Endpoint 2*), and run either simultaneously or consecutively with any number of other applications. Multipliers can also be used to represent multiple users (to a limit of 100,000 users) of each application performing concurrent or consecutive transactions. For a fuller explanation of how the endpoints operate and the agent architecture, see [Section 5.1.2.2 Passive Architecture].

Instead of using a single system simulator, I was keen to re-create the GPRS and UMTS wireless environments by using the rate-limiting feature of the testing system, Chariot[2], in an 802.11b wireless LAN. In this way, I could run the tests in a *best-case scenario*, defining accurately the exact bandwidth offered without introducing the unknown bit error factor common to the other simulators above, which attempt to re-create life-like models. This test topology removes as many unknowns from the equation and in so doing presents a

[2] Chariot™ is a trademark of NetIQ Corporation.

best-case standpoint, which has proved beneficial to me in the past when dealing with applications monitoring. From the outset, I wished to exclude such elements as:

- user traffic;
- router queuing and scheduling;
- no devices included or switched on, other than those used for testing;
- management traffic;
- serving applications to any end-user device;
- interference from other wireless LANs or wireless sources.

A diagram of the simple topology employed is given in Figure 11.1.

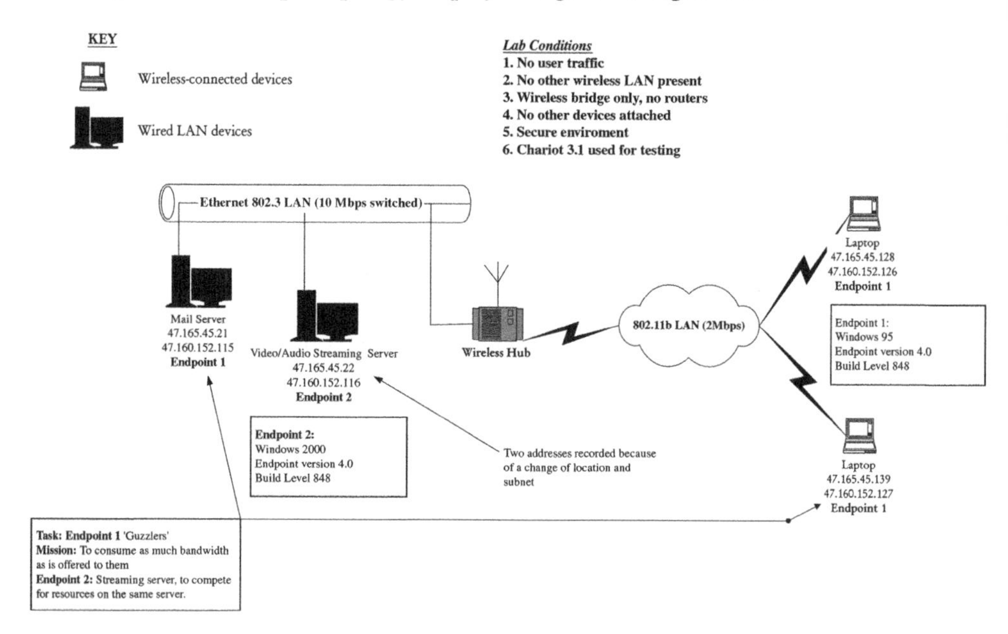

Figure 11.1 Test lab topology.

11.4 Test Results

Before settling into the tests in a wireless network in ideal conditions, it is interesting to note the results of two previous tests I gleaned from a live IP data network.

11.4.1 Live Network Results

The first[3] proves that not only is there safety in numbers, but also there are immense efficiency gains to be had from the operations of TCP, which are denied to the mobile user in a wireless network.

The transaction featured was a 'simple' Baan transaction initiated by a single key press (Tab key) resulting in no fewer than 89 network turns. The application groups involved

[3] The concurrent user test ran between Canada and the US over an ATM link shaped to 5 Mbps.

verified the Chariot script generated as being a bona fide representation of the application, which performed over restricted bandwidths in a similar way to the application itself with a 99.6% accuracy score.

Table 11.1 TCP efficiency gains

Group/ pair	Response time average	Response time minimum	Response time maximum	Response time 95% confidence interval	Measured time (s)
50 users/Pair 1	0.31460	0.31460	0.31460	n/a	3.146
100 users/Pair 1	0.32521	0.32521	0.32521	n/a	32.521
500 users/Pair 1	0.25786	0.25786	0.25786	n/a	128.932

Table 11.1 shows that a 10-fold increase in the number of users yields 57-ms efficiency, whereas a twofold increase in user numbers increases the latency, as one would perhaps expect. The 'measured time' refers to the length of time the tests took to complete.

The second series of tests was a VoIP flow without QoS applied, again on a live network, conducted at 9:00 BST across a transatlantic link[4]. There were no lost data in the stream (Test Figure 11.1).

Note that jitter is not tied directly to throughput on the link, but is independent of it. Nor does 0% data loss automatically mean that there are no issues elsewhere. In this case, a G.711 codec (at a data rate of 64 kbps) was used at a relatively quiet period while North America was still in bed, giving a very respectable jitter average of 1 ms.

This series of tests highlights in part the shift we have to make when considering QoS in relation to audio or equally to video streams. Carriers or service providers cannot think simply of response time, availability, and throughput as the defining qualities of a good service. Indeed, many have yet to reach even that stage in their SLA offerings, and those that have limit themselves to performance over their core network only. Voice quality depends on the service limiters of jitter, one-way delay, and consecutive datagram loss. Suggestions on what the limits should be vary[5], but it is true to say, as we have seen, that each can degrade the service independently of the other.

11.4.2 Laboratory Network Results

To maintain consistency across the tests the same application, Microsoft NetMeeting was used. This interactive application was chosen for its relatively low bandwidth requirements compared with other applications (both interactive and streaming). The objective was to ascertain if such an application could be used on a mobile IP device. My own experience, shared no doubt by others, told me that if a potential customer has to wait too long for the initial load sequence to complete, the product will fail to impress. The focus for me then was on the

[4] The link used for the VoIP flow was form the US to France.

[5] Values for one-way delay < 50 ms, jitter < 20 ms and lost data $< 0.2\%$ have been quoted [39], but it remains to be seen if these figures have been proven to correlate with standard mean opinion score (MOS) values for voice quality (ITU P.800) or with the more objective R ratings of the E-Model (ITU G.107).

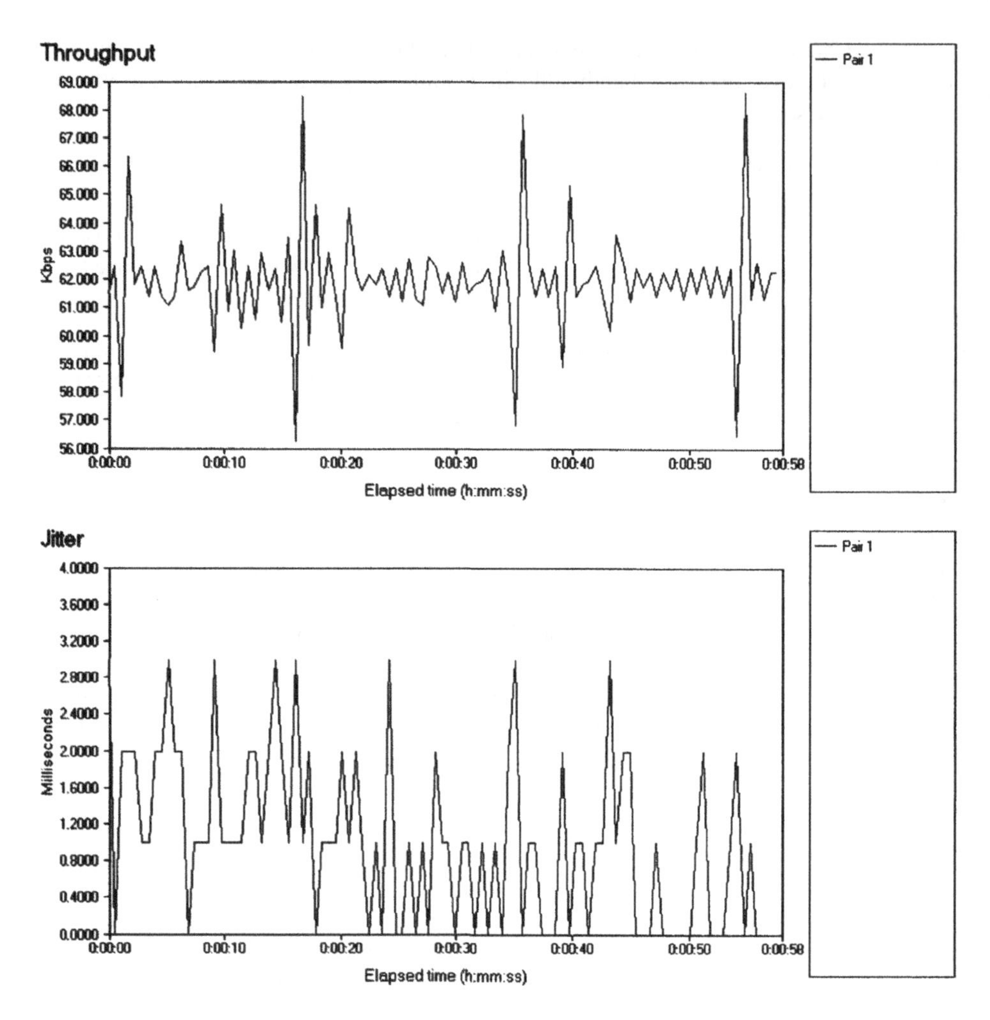

Test Figure 11.1 Jitter and throughput of a unidirectional VoIP flow.

response times for call setup rather than on the video and audio data flows that can vary widely, depending on their content and colour intensity, as described above (Section 10.10.3 Video Compression). To keep the tests as simple as possible, then, no video hardware was attached.

In the absence of any tools for measuring jitter and consecutive datagram loss, other than for VoIP, I have included *throughput measurements* to gauge the erratic behaviour of the flows when placed under stress.

The NetMeeting Call Setup was captured, the ports noted, and each TCP and UDP flow turned into an individual script, as shown in Table 11.2. These scripts were run in the sequence of the original capture concurrently between the two endpoints with the reporting carried out in *batch mode*; in other words, the tests were completed first before any reporting was done, so as to exclude any unnecessary CPU processing, which would impact on the microsecond response times recorded. No extraneous recording of CPU utilization was recorded either, for the same reason.

Table 11.2 NetMeeting call setup scripts

Group/pair	Endpoint 1	Endpoint 2	Network protocol	Service quality	Script name
All pairs					
Pair 1	47.165.45.21	47.165.45.22	TCP		NetMeeting LDAP setup_20k.scr
Pair 2	47.165.45.21	47.165.45.22	TCP		NetMeeting TCP1_Port1503_20k.scr
Pair 3	47.165.45.21	47.165.45.22	TCP		NetMeeting TCP2_Port1720_20k.scr
Pair 4	47.165.45.21	47.165.45.22	TCP		NetMeeting TCP3_Port3200_20k.scr
Pair 5	47.165.45.21	47.165.45.22	TCP		NetMeeting UDP4_Port49597_20k.scr
Pair 6	47.165.45.21	47.165.45.22	TCP		NetMeeting TCP5_Port1503_20k.scr
Pair 7	47.165.45.21	47.165.45.22	TCP		NetMeeting TCP6_Port1503_20k.scr
Pair 8	47.165.45.21	47.165.45.22	TCP		NetMeeting TCP7_Port1503_20k.scr

11.4.2.1 NetMeeting on Wired and Wireless LANs

The control test was set up between the two devices on the wired LAN to show what the best-case scenario is without restrictions on bandwidth. The throughput was between 10 and 32 kbps, and the *elapsed* time (time to complete the test) was just over 4 s (Table 11.3 and Test Figure 11.2).

Table 11.3 NetMeeting throughput rates on the control 10-Mb LAN

Group/pair	Average (kbps)	Minimum (kbps)	Maximum (kbps)	Measured time (s)
Pair 1	26.544	26.458	27.561	4.336
Pair 2	29.599	29.514	31.622	4.487
Pair 3	20.788	17.480	20.977	4.440
Pair 4	10.278	10.245	10.348	3.995
Pair 5	10.681	10.625	12.006	4.377
Pair 6	16.069	16.069	16.069	2.777
Pair 7	25.851	24.469	26.313	4.214
Pair 8	12.012	11.929	12.582	4.478

Across a wireless LAN (Test Figure 11.3), the response times begin to fluctuate but remain at around 2.5 s for pair 6, which we will focus on as it appears to be the most intensive of the flows. Throughput has decreased to 20 kbps over all the flows, and the elapsed time is now over 13 s.

Unrestricted bandwidth here means that no rate limitations were applied to the flows. As this is an 802.11b LAN, or a bridged extension of an Ethernet LAN, the full 2 Mb is not available to the application. Later, we will see that only 28% of that (or 575 kbps) is available because of the back-off protection mechanism of Ethernet LANs against congestion. For this reason, the 10-Mb LAN quoted above will only ever have 30–35% of its bandwidth available, too).

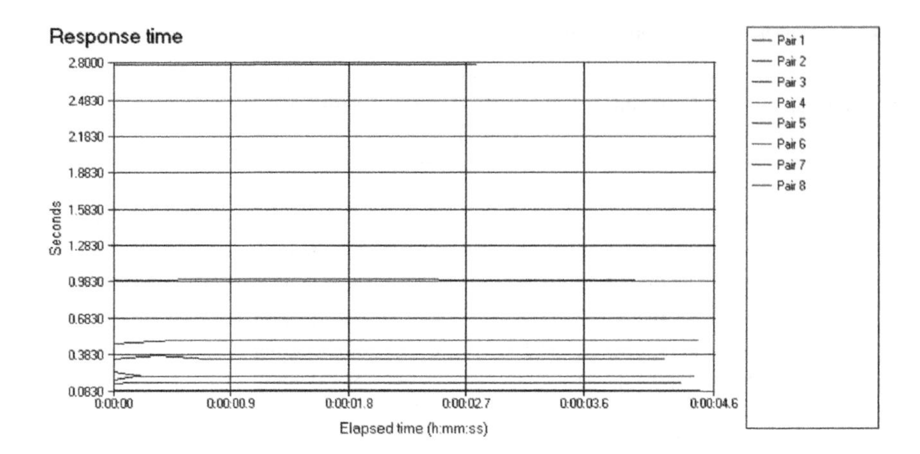

Test Figure 11.2 Control response time for a wired 10-Mb ethernet LAN.

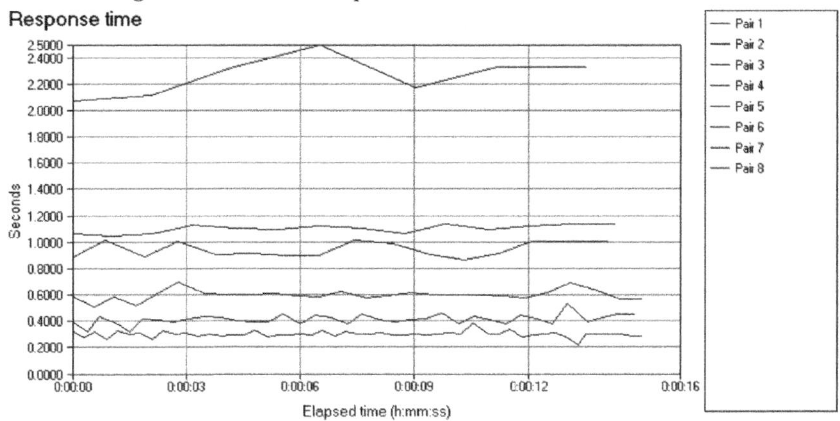

Test Figure 11.3 2-Mb wireless LAN with unrestricted bandwidth.

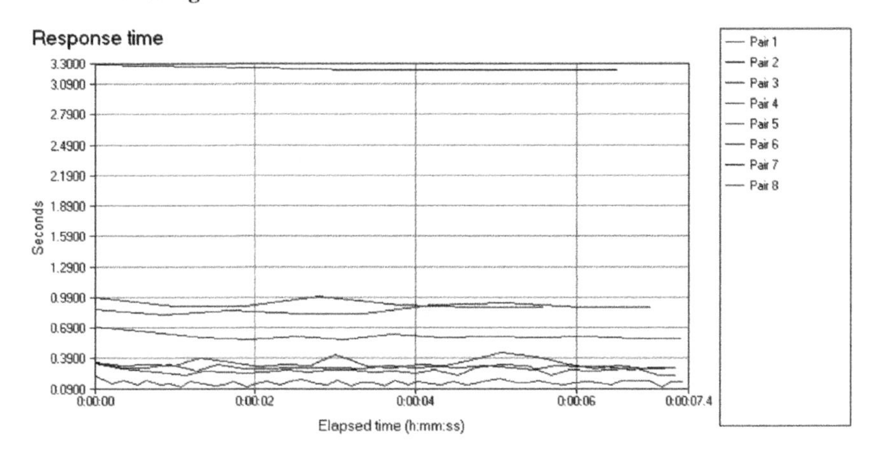

Test Figure 11.4 Response time for 20-kbps-restricted wireless LAN.

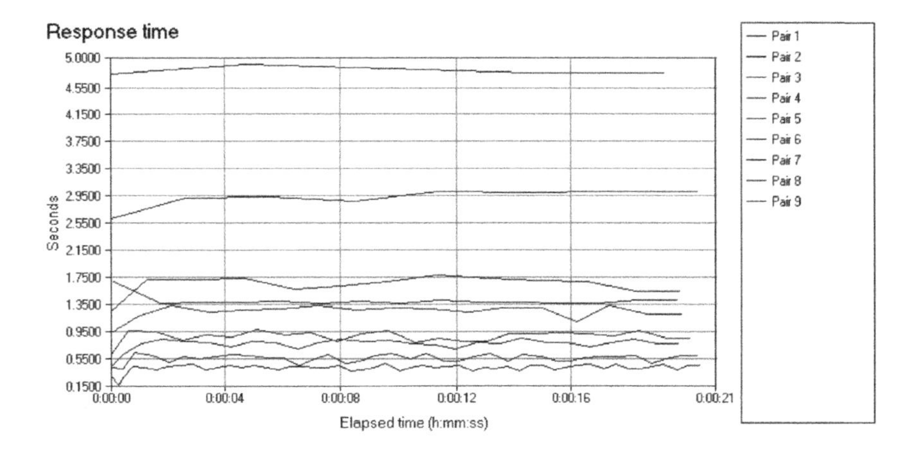

Test Figure 11.5 Response time for 20-kbps-stressed wireless LAN with guzzler.

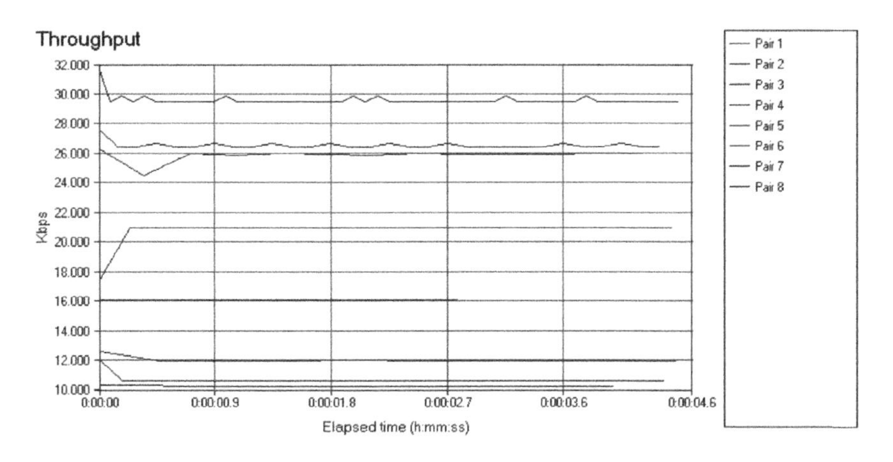

Test Figure 11.6 Control throughput for a wired 10-Mb Ethernet LAN.

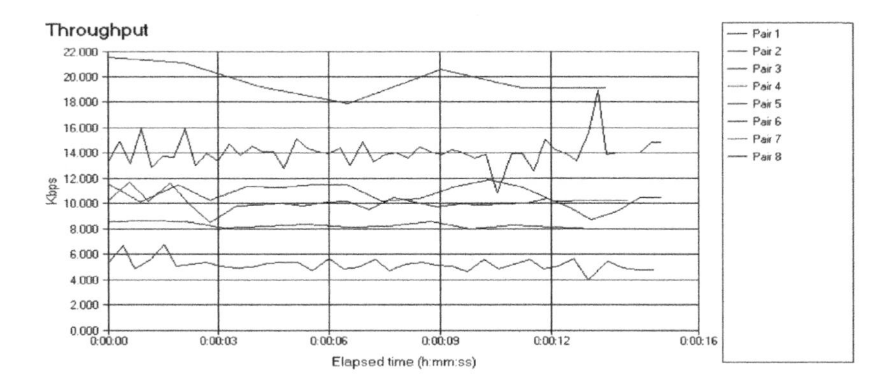

Test Figure 11.7 Throughput for 2-Mb wireless LAN with unrestricted bandwidth.

The bandwidth allocated to the application flow was then restricted to 20 kbps, using the rate-limiting feature. As can be seen in Test Figure 11.4, the response times for Pair 6 remain fairly constant but have now increased to over 3 s.

These response times are pushed out still farther to almost 5 s with the introduction of a 'guzzler' (Test Figure 11.5). This is simply a single script designed to consume all available bandwidth, which is run concurrently with the NetMeeting scripts to make them compete for bandwidth while being restricted to only 20 kbps of that bandwidth.

Throughput probably most dramatically shows how much the packet flows have been altered in moving from a wired network to an air interface. Compare Test Figures 11.6, 11.7 and 11.8 to see these effects.

Pair 2, the first TCP call, seems to have been particularly affected by the restrictions, whereas pair 6 thrives better in the 2-Mb wireless LAN than in the 10-Mb wired LAN. This underlines the point made earlier of how unreliable and meaningless throughput figures can be. Their only real value is for use with monitoring file transfer speeds.

The 'guzzler' in Test Figure 11.9 is seen to be carrying out its purpose, competing for the 20-kbps bandwidth occupied by the NetMeeting flows as well as eating all that it is permitted to by the Ethernet constraints placed upon it.

To better appreciate and compare the relative performances of pair 6 in different scenarios, Tables 11.4 and 11.5 show additional test results for increasing amounts of restricted bandwidth under the stress of the guzzler.

Table 11.4 Pair 6 response time results for multiple bandwidths

Response timepair 6 results	Response time average	Response time minimum	Response time maximum	Measured time (s)
10 Mbps unrestricted	2.77700	2.77700	2.77700	2.777
2 Mbps unrestricted	2.25183	2.07400	2.49400	13.511
20 kbps restricted	3.26100	3.23400	3.28800	6.522
20 kbps stressed	4.81250	4.74700	4.90200	19.250
56 kbps stressed	4.14625	4.00000	4.38000	16.585
64 kbps stressed	4.54925	3.73100	6.02400	18.197
128 kbps stressed	4.28500	4.13100	4.51000	12.855

Table 11.5 Pair 6 throughput results for multiple bandwidths

Throughput pair 6 results	Average (kbps)	Minimum (kbps)	Maximum (kbps)	Measured time (s)
10 Mbps unrestricted	16.069	16.069	16.069	2.777
2 Mbps unrestricted	19.817	17.893	21.516	13.511
20 kbps restricted	13.684	13.572	13.799	6.522
20 kbps stressed	9.273	9.103	9.401	19.250
56 kbps stressed	10.763	10.188	11.156	16.585
64 kbps stressed	9.809	7.408	11.961	18.197
128 kbps stressed	10.414	9.895	10.802	12.855

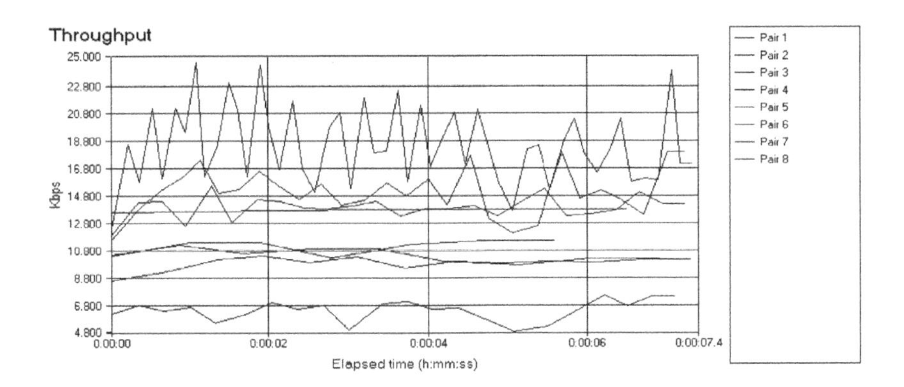

Test Figure 11.8 Throughput for 20-kbps-restricted wireless LAN.

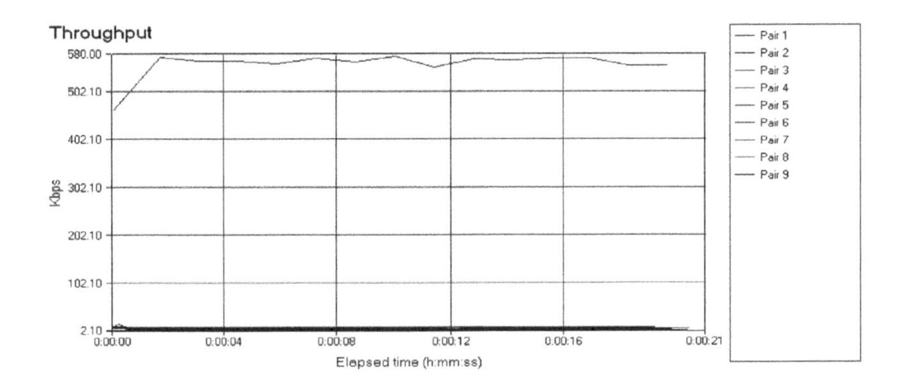

Test Figure 11.9 Throughput for 20-kbps-stressed wireless LAN with guzzler.

It is clear that response times under stress conditions for this part of the NetMeeting application flow are not going to fare any better with increased amounts of bandwidth, while in the presence of fierce competition for its own share of that bandwidth, than the 4 s shown above. This is good and bad news.

11.4.2.2 GPRS Implications

When transposed to a GPRS network operating with coding scheme CS-4, this would indicate that 4 s is as long as it would take to set up the NetMeeting call, which is not an eternity. Unfortunately, it would also appear that no amount of increased bandwidth in the shape of additional timeslots or *multislots* will make it any faster, when other applications are competing against it. This limitation may manifest itself as a restriction in the number of applications permitted to be open at any one time, reminiscent of earlier PC operating systems (and some much more recent operating systems, too). Perhaps we can look forward to the time when mobile IP devices have multithreaded operating systems to cope with such demands.

Tables 11.4 and 11.5 extend over and above what GPRS is capable of now and even what it or EGPRS may be capable of in the future (Section 9.3.7.5 Mean Throughput Class). Remember that EGPRS extends only up to 59.2 kbps, GPRS can only realistically manage 40 kbps, assuming that two timeslots (MSC2) are used (and not all users within a cell demand it), and the highest mean throughput value is 111 kbps. The values stated at 64 kbps and above could apply equally to the third generation of networks.

11.4.2.3 UMTS Implications

While the war of words continues over how much bandwidth will be delivered in 3G networks from day 1, it seems reasonable to assume that a minimum of 64 kbps and a premium service of a maximum of 384 kbps will be available within the first two years of service rollout. With this in mind, Table 11.6 represents a possible UMTS scenario at the premium level where a video streaming or playback video application under stress is compared with an unstressed audio stream. Both are non-real-time, and therefore, one would that assume jitter-buffering mechanisms would be available in mobile devices to cope with any undesired delay variations.

Table 11.6 Video and audio stream response times and throughput

Response time results	Response time average	Response time minimum	Response time maximum	Measured time (s)
Video stream over 384 kbps stressed	8.36142	8.17300	10.41900	418.071
Audio stream unrestricted	0.82214	0.36200	9.16900	41.107
Throughput results	**Average (kbps)**	**Minimum (kbps)**	**Maximum (kbps)**	**Measured Time (s)**
Video stream over 384 kbps stressed	370.995	297.730	379.548	418.071
Audio stream unrestricted	469.278	42.078	1,065.780	41.107

'Stressed' here and in the preceding tables includes the application flows being restricted to their allocated bandwidth. There is an erratic element at the end of the audio stream, which is explained by the inordinately long time that it took to cease the stream and tear down the connection. The averages should be taken as representative of the majority of the flow.

Two things are striking about these results: first, even under stress, the video streaming application seems better able to cope given the additional bandwidth and defends its allocation very well with a credible 370 kbps out of a possible 384 kbps. Second, and not surprisingly, the audio stream demands are much less stringent and the response times much healthier as a result. This leads me to believe that music will be the first application we will hear from 3G handsets before video. The quality and synchronization of the picture will need to compensate for the 8-s delay, so that the failure of video-conferencing and desktop cameras to move from the realms of quirky toy to mass adoption is not repeated.

11.4.2.4 File-transfer Implications

In this test, a data rate of 28.8 kbps was chosen, as not only is it a recognized dial-up modem rate, with which most of us can identify and therefore make reasonable comparisons, but also it equates to just over twice the expected data rate on a *GPRS network using CS-2 and a multislot capability of two timeslots*. It will be remembered that coding scheme two (CS-2) was the recommended scheme for normal radio conditions from the GPRS simulator report and also that CS-2 returned a data rate for one timeslot of 13.4 kbps (Section 9.3.5 Coding Schemes).

The response times and throughput were constant, as no other traffic occupied the wireless LAN and the FTP script represented both control and data flows. One can conclude from this that in an error-free and non-competitive wireless network, at least, transfers deliver better throughput than 28.8 kbps modems and compare favourably with the current throughput performance of a 56-kbps modem, which struggles to break the 30 kbps throughput barrier most days. The wait of over four and a half minutes, however, is going to deter the use of e-mail with large attachments or the downloading of large music files in a GPRS network (Table 11.7).

Table 11.7 FTP transfer of 1 Mb for 28.8 kbps for a restricted wireless LAN

Group/ pair	Response time average	Throughput average (kbps)	Bytes sent by E1	Bytes received by E1	Measured time (s)
Pair 1	277.83900	28.123	1,000,060	90	277.839

11.4.2.5 Videophone Possibilities

There have been prototypes of this type of technology already demonstrated, and it is interesting to ascertain what sort of bandwidths are likely to be necessary, regardless of the QoS parameters that would need to be set to guarantee a service level. Table 11.8 shows how NetMeeting locates another party ready for interactive audio and video to begin.

This is the natural progression from the NetMeeting call setup studied above. Once the application has been loaded, you would find a number in your address book similar to a telephone directory entry today and click on the Call button. The breakdown above is what happens after you press Call and the called party is live or has their IP device turned on. The located party in this case was in Canada with the calling party in the UK across a live corporate intranet.

With the largest flow of 51k passing from the called client, the expected delay in 'connecting' would be as much as 20 s over a 20-kbps link, returning to our GPRS, CS-4/MSC1 scenario. In order to arouse some enthusiasm in your potential market, you would need to offer between 64 kbps and an optimum 128 kbps to connect in 3–6 s. At between 4–5 Mbps, the transatlantic link provided a response in less than 23 s, which would suggest that international calls might be a distinct possibility on interactive videophones, if sufficient bandwidth could be provided to each subscriber so as not to lengthen that response to more than 30 s in total. Most I think would be willing to wait that long to see and speak to their families on the other side of the globe.

Table 11.8 NetMeeting locating called party and called party's response

Stream	Server or called client	Server or called client port number	Bytes sent	Bytes received	Calling client port number	Transport protocol	Response time (ms)
Pair 1	LDAP server	389 (LDAP)	162	129	1299	TCP	318
Pair 2	Called client	1503	202	130	NA	TCP	233
Pair 3	Called client	1720	413	124	NA	TCP	1
Pair 4	Called client	3177	604	640	NA	TCP	546
Pair 5	Called client	1503	755	411	NA	TCP	2814
Pair 6	Called client	1503	3 731	2 559	NA	TCP	17,994
Pair 7	Called client	1503	45 110	1 944	NA	TCP	881
Totals			50 977	5 937			22,787

11.4.2.6 Choose a Colour

As a last thought, we discussed how different methods are used to compress images (Section 10.10.3 Video Compression). In the course of testing NetMeeting I examined the interactive whiteboard facility to see how it behaves in a dormant state and when a circle is drawn and filled with a colour to determine if colour choice would have any noticeable impact on its potential use across wireless networks. All bytes recorded are pure payload bytes with headers removed and the colours used are 8-bit (256 colours) rather than the true colour 32-bit (16 million colours), which increases the size and bandwidth required threefold.

We discovered earlier that monotone colours compress easily. The results in Table 11.9 indicate a good deal of variation, with black being the most economic colour in terms of bytes sent and received, and yellow being the most expensive. Also, it is worth noting that anyone contemplating running whiteboard applications as part of their wireless service offering would need to budget for over 63 bytes per second in idle mode, unless some kind of spoofing were employed.

Table 11.9 Colour choice in NetMeeting's whiteboard facility

State or colour	Server or called client port number	Bytes sent	Bytes received	Net[a] bytes sent	Net bytes received	Elapsed time (s)
Dormant	49600	2844	0	63.2 bps	NA	45
Circle – no fill	49601	975	130	88	108	10
Green fill	49601	1023	126	28	104	6
Black fill	49601	2519	74	48	52	4
Magenta fill	49601	963	82	76	60	8
Yellow fill	49601	2843	126	48	104	5
TCP confirm – common to all	1503	NA	NA	131	22	NA

[a] Net refers to bytes sent or received less the cost of lying dormant.

The reticence in taking up wireless application protocol (WAP) services was in part due to its monotone screen and low-resolution display. Having seen the full colour displays now available, customers are understandably more inclined to wait for the bright, multicoloured future to arrive at an affordable price.

12

Conclusions

This chapter presents a synopsis of the test results, a summary of what elements will contribute towards guaranteeing service levels when applications cross the air interface from fixed to wireless networks, and what issues will need to be addressed to ensure that using mobile IP devices as a portal to services does not disappoint.

12.1 Synopsis of Test Results

The tests were designed to encompass a number of potential application types that GPRS and UMTS networks may be expected to support. Based on previous work, agent architecture was employed to report on real-time application performance across a real air interface, rather than rely on constructed models and protocol stacks built on a single system. I believe both to be complementary, as there is a distinct lack of empirical data to support the deployment of applications, new or old, into 2.5G and 3G networks. With this in mind, I concluded the following from the tests:

- As the numbers of users increase in a TCP-controlled flow in a fixed network, efficiency gains are to be had. A twofold increase yields a higher latency, but a 10-fold increase yielded a 57 ms reduction in latency.
- Jitter works independently of throughput, and bad or erratic throughput does not necessarily degrade VoIP performance.
- Limiting the data rate to a single timeslot, best-case GPRS scenario increased latency marginally.
- Limiting the data rate and introducing a competing application (under stress conditions) doubled the latency over the original unrestricted bandwidth scenario.
- Throughput is not a sound yardstick on which to base SLAs except for file data transfers.
- A NetMeeting call would take no more than 4 seconds to set up.
- No amount of additional bandwidth up to the theoretical limits of GPRS and EGPRS will improve on the 4-seconds NetMeeting call setup time under stress conditions.
- Lower rates of 64 kbps and 128 kbps in a 3G network will experience the same limited call setup time.
- Video streaming applications would be able to cope with stress conditions in a 3G network when allocated the premium rate of 384 kbps.
- FTP transfers over what resembles a GPRS scenario operating under normal radio conditions, using two timeslots, compares favourably with the performance of a 56-kbps dial-up

modem. The four and a half minutes to download a 1-Mb file will preclude e-mail with attachments and large music file downloads from being of practical benefit in a GPRS network.

- Connecting via videophone would be optimized at 128 kbps per subscriber to ensure a connection time of 3 seconds.
- Lower-resolution colour depth and, in some cases, individual colours can be more economical in their use of bandwidth.
- Wireless networks wishing to exploit NetMeeting's interactive whiteboard application will need to budget 63 bytes per second in its 'dormant' state.

12.2 Summary

In the fixed or wired data network world, we have been used to low bit error rates, high throughput across the network cores and reasonable dial-up access speeds up to and above 128 kbps. Hardware failures, packet loss through congestion, slow re-convergence of routes, and construction engineers have been their public enemies.

In the wireless voice and data world and with wireless IP devices set to proliferate in the coming years, available bandwidth is low, especially when many share a cell's resources, bit error rates are high, and jitter and delay are commonplace. Signal encoding, compression and forward error correction all play their part in contributing to those characteristics. What is certain is that service performance and the guarantees that go with it are critical to the success and acceptance of mobile devices becoming the portal to multiservice networks.

Much of the continuing work focuses quite rightly on the mechanisms and protocols necessary to classify packets or flows, assign correct levels of QoS, set up a path to guarantee that QoS end to end, and speed them on their way. The nuts and bolts of router queuing and scheduling mechanisms, packet classification, minimizing interference, coding schemes, mobility and connection management are necessary components of what could be termed a multiservice architecture to ease their passage from fixed to wireless.

What will become the glue of that multiservice architecture will be the simple but flexible schemes that have monolithic consequences. We have two already in MPLS and IPv6, and there may be more to follow.

IPv6 was an opportunity to simplify the unwieldy header that IPv4 had evolved, increase the number of available device addresses well into this millennium, and make modest additions to its format. Now that it is a mandatory element of the UMTS architecture, it is an opportunity to clean up the IP act in readiness for what lies ahead.

MPLS not only offers a middle road between the IP and ATM camps, but also more importantly provides a simple, extensible transport mechanism that can be pushed out to the very edges of the radio access network, can draw on the strengths of both circuit-switched and packet-based networks to route around problems with minimum fuss (to give one example), and can be adapted to suit future technologies.

It is somewhat ironic that neither of these defines quality of service. They merely carry it, but they also help to maintain it, which in turn helps to guarantee it. None of the QoS traffic classes, QoS, COS, or IP TOS elements do that. Defining is essential, but guaranteeing is crucial.

We have seen earlier with agent technology how it is possible to track real-time application performance, which is the only true measure of what a customer is feeling *right now*. This helps us not only to understand the inevitable frustrations felt during slowdowns in performance, but also to be truly proactive in the day-to-day care of customers, knowing that one can be responsible for avoiding a potentially damaging event for one's business by homing in on the root cause quickly and effectively.

The challenge for the companies that gave us agents is to adapt them to wireless environments in order to capture critical components that put service levels at risk, such as monitoring reduced coverage in a cell, bandwidth allocations in terms of both requested and received quality of service, and power fluctuations from the terminal.

Looking to the future, we also touched on admission policies to accept traffic into the network, and this points to another essential element, which has yet to be defined fully, that would control the end-to-end management of the resources and flows. In my opinion, the crucial element that would feed a higher pedigree of policy manager (which could be called a multiservice manager) with the current change status of the network would be a stateful[1], application-level aware manager, intelligent enough to monitor exceptions to the *status quo* rather than every single flow. That is deserving of further study outside the scope of this book.

For now, it is enough to know that technologies are expanding and stretching the finest minds to bring us image and knowledge capabilities that have hitherto had us glued to our sofas. Fears expressed about the anti-social nature of Baird's invention when it first came on to the market resurface with each new piece of mobile equipment, but in my mind, a device that brings me the faces as well as voices of some of my dearest friends from half way round the globe is worth every dollar, pound, or euro.

The challenge for those of us that work in the industry is to make that sure it works consistently well, is successful, and exceeds expectations, where possible. The management and attention paid to service levels are key to that success. I look forward to the challenge.

[1] A stateful system is one that can adapt and react immediately to dynamically changing circumstances. Monitoring traffic flows is the bread and butter of these systems. An application-level aware firewall is one such system.

References

1. Case J.D., Fedor M.S., Schoffstall M.L., Davin J.R., Simple Network Management Protocol RFC 1157, May 1990.
2. Case J.D., Fedor M.S., Schoffstall M.L., Davin J.R., A Simple Gateway Monitoring Protocol RFC 1028, November 1987.
3. Cerf V., IAB Recommendations for the Development of Internet Network Management Standards RFC 1052, April 1988.
4. Rose M., McCloghrie K., Structure and Identification of Management Information for TCP-IP Based Internets RFC 1155, May 1990.
5. McCloghrie K., Rose M., Management Information Base for Network Management of TCP-IP Based Internets MIB-II RFC 1213, March 1991.
6. Stallings W., Network Management IEEE Computer Society Press, 1993.
7. Rabie S., Dam Xi-Nam, 'An Integrated Architecture for LAN/WAN Management'. In Proceedings of the IEEE Network Operations and Management Symposium, 1992, pp. 254265.
8. Cerf V., Report of the Second Ad Hoc Network Management Review Group RFC 1109, August 1989.
9. Stallings W., SNMP, SNMPv2, SNMPv3 and RMON1 and 2, Addison Wesley Longman, 1999.
10. OSI, I.S.O. 10040 Systems Management Overview, 1991.
11. Rose M.T., The Open Book, A Practical Perspective on OSI, Prentice Hall, Englewood Cliffs, NJ, 1990.
12. Yemeni Y., 'The OSI Network Management Model', IEEE Communications Magazine, May 1993, pp. 2029.
13. Hayes S., Analysing Network Performance Management, IEEE Communications Magazine, May 1993, Vol. 31, Number 5.
14. Gartner Group Applications Development & Management Strategies Note, February 2000, COM–03–8396.
15. Sturm R., 'Need for End to End Management', In Perspectives Reseaux 1997.
16. DMTF Common Information Model (CIM) Specification, v2.0, March 1998.
17. DMTF Application Model White Paper, Draft v0.9, May 1998.
18. Adams A., Mahdavi J., Mathis M., Paxson V., 'Creating a Scalable Architecture for Internet Measurement', In Proceedings INET 98, Geneva, July 1998.
19. Nemzow M.A.W., Enterprise Network Performance Optimization, McGraw-Hill, New York, 1994.
20. Digital Cellular Telecommunications System (Phase 2 +), General Packet Radio Service, Service Description, Stage 1, GSM 02.60 v6.2.1. August 1999.
21. Croll A., Packman E., Managing Bandwidth Deploying QoS in Enterprise Networks. Prentice Hall, Englewood Cliffs, NJ, 2000.

22. Zipperle A., Using Chariot to Validate Prioritisation of VoIP Traffic Application Note, November 2000.
23. Digital Cellular Telecommunications System (Phase 2 +), General Packet Radio Service (GPRS), Service Description, Stage 2, GSM 03.60 v6.3.1. Release 1997.
24. Digital Cellular Telecommunications System (Phase 2 +), General Packet Radio Service (GPRS), GPRS Tunnel Protocol Across the Gn and Gp Interface, GSM 09.60 v6.2.2. Nov 1998.
25. Digital Cellular Telecommunications System (Phase 2 +), General Packet Radio Service, Subnetwork Dependent Convergence Protocol, GSM 04.65 v6.3.0. March 1999.
26. Digital Cellular Telecommunications System (Phase 2 +), General Packet Radio Service, Overall Description of the GPRS Radio Interface, GSM 03.64 v8.6.0. September 2000.
27. Digital Cellular Telecommunications System (Phase 2 +) General Packet Radio Service (GPRS) Service Description Stage, GSM 03.60 Version 7.6.0 Release 1998.
28. Kalden R., Meirick I., Meyer M., 'Wireless Internet Access Based on GPRS. Aachen, Germany Ericsson Eurolab Deutschland'. In IEEE Personal Communications, April 2000.
29. Dixit S., Guo Y., Antoniou Z., 'Resource Management and Quality of Service in Third-generation Wireless Networks. Burlington, MA Nokia Research Center', IEEE Communications Magazine, February 2001.
30. Braden R., Clark D., Crowcroft J., et al., Recommendations on Queue Management and Congestion Avoidance in the Internet Informational RFC 2309, April 1998.
31. Huston G., Internet Performance Survival Guide QoS Strategies for Multiservice Network, John Wiley, New York, 2000.
32. Kleinrock L., Queuing Systems, Vol. 2 Computer Applications, John Wiley, New York, 1976.
33. Black U., QoS in Wide Area Networks. Prentice Hall, Englewood Cliffs, NJ, 2000.
34. Jacobson V., 'Congestion Avoidance and Control', Computer Communication Review, vol. 18, no. 4, August 1988.
35. Stevens W.R., TCP/IP Illustrated, Vol. 1. Addison-Wesley, Reading, MA, 1994.
36. Davie B., Rekhter Y., MPLS Technology and Applications. Academic Press, Morgan Kaufmann, 2000.
37. Li T., Rekhter Y., A Provider Architecture for Differentiated Services and Traffic Engineering (PASTE) RFC 2430 October 1998.
38. Gruhl S., Echihabi A., Rachidi T., Link M., Sollner M., A Demonstrator for Real-time Multimedia Sessions over 3rd Generation Wireless Networks Nuremburg, Germany and Ifrane, Morocco Bell Laboratories, Lucent and Alakhawayn University 2000 IEEE 0–7803–6536–4/00.
39. Walker J.Q., Hicks J., Using Chariot to evaluate data networks for Voice Readiness White paper, NetIQ Corporation, 2001.

Appendix A

The measurements were taken across the same WAN link every 10 min at the same time of day for both PING and, in this case, a simulated Baan transaction.

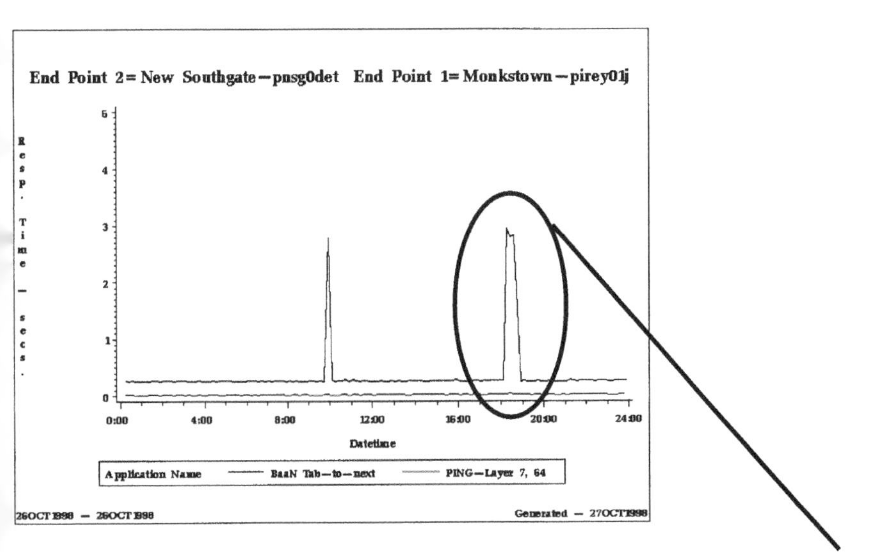

Here, the synthetic transaction "BaaN Tab-to-Next" detected service degradation. Note that the "Ping" transaction did not detect this degradation.

Figure A.1 Two opposing views of application performance degradation.

Synthetic transactions are used to model a real transaction on the network thereby exluding the application from the measurement.

Appendix B

An example of part of the CIM Core Model showing the Management System Element class and one of its subclasses, Product.

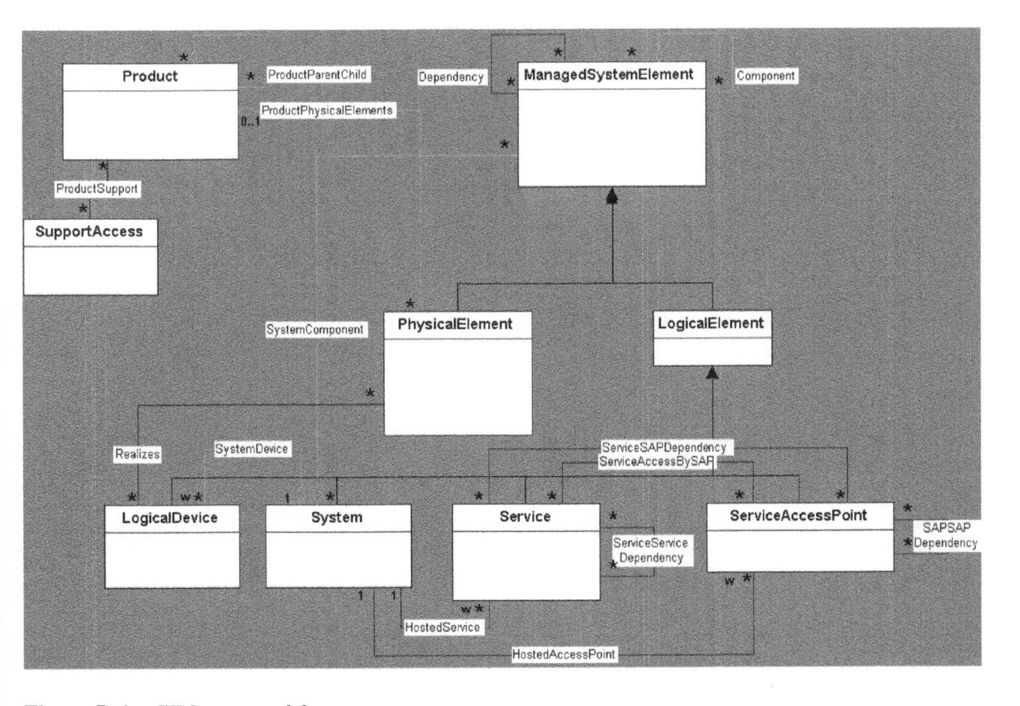

Figure B.1 CIM core model extract.
Source dMTf Core Model White Paper v1.0 August 5th, 1998.

Appendix C

Diffserv code points (DSCPs), known sometimes as DS fields and IP TOS fields, use the same 8 bits of the IP header but in different ways, as set out below. While the IP precedence field is still present, albeit in a different position, it provides some level of backwards compatibility. The Type of Service field, however, would need translating or 're-marking' to match a DS field. Note that with DS code points, the least significant bit is on the left.

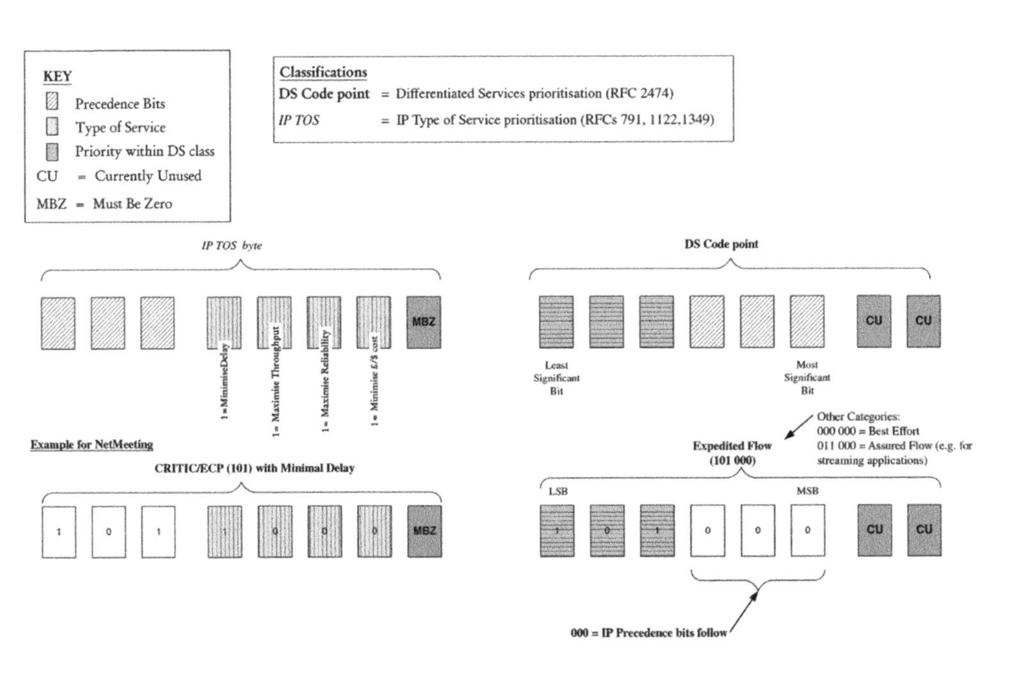

Figure C.1 IP TOS and DSCP fields.

Microsoft releases of Windows[1] 98 and Windows 2000 with QoS Packet Scheduler enabled defines 3 DSCPs in their Generic quality of service (GQOS). They are:

000 000 Best Effort
011 000 Controlled Load
101 000 Guarantee

[1] Windows™ is a trademark of Microsoft Corporation.

Appendix D

The IPv6 header grows the address field from 32 to 128 bits, giving a possible 2^{128} IP addresses, envisaging a time when each person on the planet has over 100 IP devices connected with him or her. Other features include:

- Security provided through inherent support for IPSec
- Flow field introduction to enhance QoS capabilities within flows
- Auto-configuration of addresses for mobile IP devices

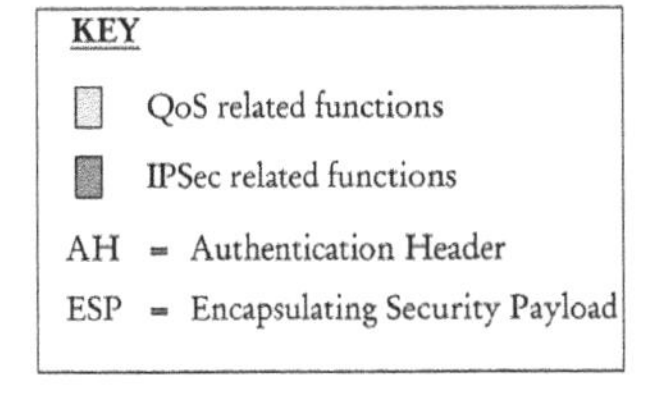

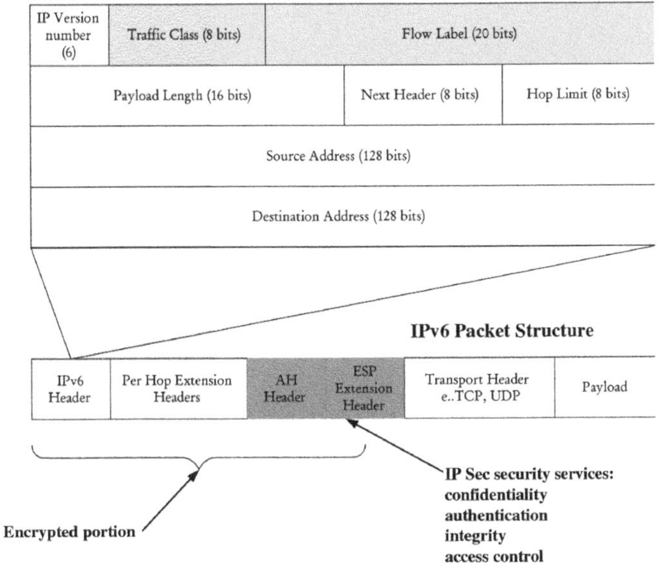

Figure D.1 IPv6 header and structure.

Appendix E

The MPLS label is a simple 32-bit number broken down into the fields shown below.

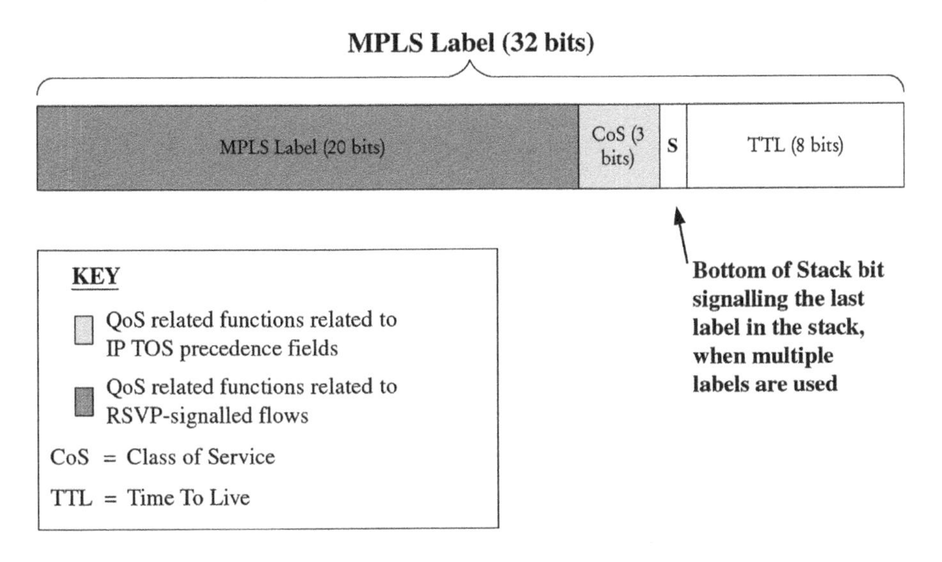

Figure E.1 MPLS label format

Abbreviations

3GPP	3rd Generation Partnership Project
AF	Assured Forwarding
AGCH	Access Grant Channel
ARM	Applications Response Measurement
AS	Autonomous System
ASK	Amplitude Shift Keying
ASN	Abstract Syntax Notation
AuC	Authentication Centre
BCCH	Broadcast Control Channel
BCS	Block Check Sequence
BER	Bit Error Rate
BGP4	Border Gateway Protocol version 4
BSC	Base Station Controller
BSS	Base Station Subsystem
BTS	Base Transceiver Station
CDMA	Code Division Multiple Access
CIDR	Classless Inter-Domain Routing
CIM	Common Information Model
CM	Communications Management
CMIP	Communications Management Information Protocol
CMIS	Communications Management Information Services
CMOT	CMIP over TCP/IP
COS/CoS	Class of Service
CS	Coding Scheme
DCS	Digital Cellular System
DHCP	Dynamic Host Configuration Protocol
DiffServ	Differentiated Services
DMTF	Desktop Management Task Force
DNS	Domain Name Service

DQPSK	Differential Quadrature Phase Shift Keying
DRX	Discontinuous Reception
DSCP	DiffServ Code Point (8-bit field adapted from IP TOS field, 6-bit code point including 3-bit Priority, 3-bit Precedence field and 2 unused bits). See Appendix C.
DTM	Dual Transfer Mode
DTX	Discontinuous Transmission
ECSD	Enhanced Circuit-Switched Data
EDGE	Enhanced Date rate for Global Evolution
EF	Expedited Forwarding
EGPRS	Enhanced General Packet Radio Service
EIR	Equipment Identity Register
ERP	Enterprise Resource Planning
FCCH	Frequency Control Channel
FDMA	Frequency Division Multiple Access
FEC	Forward Error Correction
FSK	Frequency Shift Keying
FTP	File Transfer Protocol
GERAN	GPRS/Edge Radio Access Network
GGSN	Gateway GPRS Support Node
GIF	Graphics Interchange Format
GMPLS	Generalised MPLS
GMSK	Gaussian Minimum Shift Keying
GPRS	General Packet Radio Service
GPS	Global Positioning Systems
GQOS	Generic Quality Of Service
GSM	Global System for Mobile communications
GTP	GPRS Tunnel Protocol
HDR	Higher Data Rate (U.S. standard only)
HLR	Home Location Register
HSCSD	High Speed Circuit Switched Data
IAB	Internet Activities Board
ICMP	Internet Control Message Protocol
IGMP	Internet Group Management Protocol
IMEI	International Mobile Equipment Identity
IMS	IP Multimedia System
IMSI	International Mobile Subscriber Identity
IntServ	Integrated Services
IPX	Internet Packet eXchange
ISDN	Integrated Services Digital Network
ISO	International Standards Organisation

ITU	International Telecommunications Union
JPEG	Joint Photographic Experts Group
LAN	Local Area Network
LLC	Logical Link Control
LMAE	Local Management Application Entity
LSR	Label Switching Router
MAC	Media Access Control
MAC (GPRS)	Medium Access Control
MAHO	Mobile Assisted Hand-Off
MCS	Modulation and Coding Scheme (EGPRS only)
MIB	Management Information Base
MIT	Management Information Tree
MM	Mobility Management
MOF	Managed Object Files
MOS	Mean Opinion Score
MPLS	Multi-Protocol Label Switching
MS	Mobile Station
MSC	Mobile services Switching Centre
MSK	Minimum Shift Keying
MTSO	Mobile Telephone Switching Office
MTU	Maximum Transmission Unit
MWIF	Mobile Wireless Internet Forum
NIC	Network Interface Card
NIMI	National Internet Measurement Infrastructure
NMS	Network Management Station
NSS	Network Subsystem
NTP	Network Time Protocol
OO	Object Oriented
OSI	Open Systems Interconnection
OSPF	Open Shortest Path First (where 'open' refers to the open standard that it is)
PACCH	Packet Associated Control Channel
PAGCH	Packet Access Grant Channel
PBCCH	Packet Broadcast Control Channel
PCCCH	Packet Common Control Channels
PCH	Paging Channel
PCMCIA	Personal Computer Memory Card International Association
PCN	Personal Communications Network
PCS	Personal Communications Service
PDCH	Packet Data Channel
PDP	Packet Data Protocol

PDTCH	Packet Data Transfer Channel
PDU	Protocol Data Unit
PHB	Per Hop Behaviour
PING	Packet InterNet Groper
PPCH	Packet Paging Channel
PPP	Point-to-Point Protocol
PRACH	Packet Random Access Channel
PSK	Phase Shift Keying
PSTN	Public Switched Telephone Network
QoS	Quality of Service
RACH	Random Access Channel
RDN	Relative Distinguishing Name
RED	Random Early Detection
RLA	Received Level Average
RLC	Radio Link Control
RMON	Remote Network Monitoring
RPE-LPC	Regular Pulse Excited Linear Predictive Coder
RR	Radio Resource
RSVP	Resource reSerVation Protocol
RTCP	Real Time Control Protocol
RTP	Real Time Protocol
RTT	Round Trip Time
SCH	Synchronisation Channel
SGSN	Serving GPRS Support Node
SIM	Subscriber Identity Module
SLA	Service Level Agreement
SMI	Structure of Management Information
SMS	Systems Management System
SMS (Wireless)	Short Message Service
SNMP	Simple Network Management Protocol
SR-ARQ	Selective Repeat Automatic Repeat Request
TBF	Temporary Block Flow
TCA	Traffic Conditioning Agreement
TCH	Traffic Channels
TCP	Transmission Control Protocol
TDMA	Time Division Multiple Access
TOS	Type Of Service (8-bit field, comprising 3-bit Precedence field, 4-bit TOS and 0-bit)
Tr-TCM	Two-rate, Three-Colour Marker
UDP	User Datagram Protocol
UMTS	Universal Mobile Telecommunications System

USM	User-based Security Model
UTRAN	UMTS Terrestrial Radio Access Network
VACM	View-based Access Control Model
VLAN	Virtual LAN
VLR	Visitor Location Register
VoIP	Voice over Internet Protocol
VPN	Virtual Private Network
WAN	Wide Area Network
WAP	Wireless Application Protocol
WBEM	Web-Based Enterprise Management
W-CDMA	Wideband Code Division Multiple Access
WLAN	Wireless LANs (802.11b specification)

Index

(Figures in italics)

Printed and bound in the UK by
CPI Antony Rowe, Eastbourne

Printed and bound by CPI Group (UK) Ltd, Croydon, CR0 4YY

23/04/2025

14660945-0002